Math 119 College Algebra

Note-Taking Guides

Updated August 2025

Cengage

Australia • Brazil • Canada • Mexico • Singapore • United Kingdom • United States

Cengage

Math 119 College Algebra, Note-Taking Guides

Math 119 College Algebra: Note-Taking Guides
Angela Gentry, Jack Ryan, April Shilling-Lambert, Kim Smith, and Shel Swenson

WCN: 01-100-496

ISBN: 979-8-214-53477-0

Cengage
10650 Toebben Drive, Suite A
Independence, KY 41051
USA

Cengage is a leading provider of customized learning solutions with employees residing in nearly 40 different countries and sales in more than 125 countries around the world. Find your local representative at: **www.cengage.com.**

To learn more about Cengage platforms and services, register or access your online learning solution, or purchase materials for your course, visit **www.cengage.com**.

Math 119 - College Algebra

Note-Taking Guides

Table of Contents

Unit 1

Fundamental Concepts of Algebra

This unit lays the foundation for understanding and applying key algebraic concepts, such as real numbers, integer exponents, radical exponents, rational exponents, polynomial expressions, rational expressions, and factoring. By the end of the unit, students will have the algebraic building blocks, ready to tackle the more advanced topics in the following units.

NAME: ______________________________ EMAIL: ______________________ @vols.utk.edu

Lesson 1: Real Numbers

Learning Objectives:

I. Classify sets of real numbers as natural numbers, integers, rational numbers, and/or irrational numbers.

II. Order real numbers using the real number line and inequalities. Write sets of real numbers in interval and inequality notation.

III. Evaluate expressions using the order of operations.

I. Sets of Real Numbers (Natural, Whole, Integer, Rational, Irrational, and Real)

A ______________________ is the formal term in mathematics that is used to refer to a collection of objects. For example, the set $\{1, 2, 3\}$ contains

Note: We always use a pair of braces $\{\ \}$ to enclose the members of a set.

A set A is called a ______________________ of a set B if every member of A is also a member of B.

For example,

- $\{1, 2, 3\}$ ______________________ a subset of $\{1, 2, 3, 4\}$.
- $\{0, 4\}$ ______________________ a subset of $\{0, 1, 2, 3, 4\}$.
- $\{0, 5\}$ ______________________ a subset of $\{0, 1, 2, 3, 4\}$.

There are a variety of types of numbers that make up the real number system.

- Natural Numbers:

- Whole Numbers:

- Integers:

- Rational Numbers:

- Irrational Numbers:

The set of all real numbers is denoted by the symbol $\mathbb{R}$. Let's draw a diagram to help us classify the types of real numbers that we will encounter in this class.

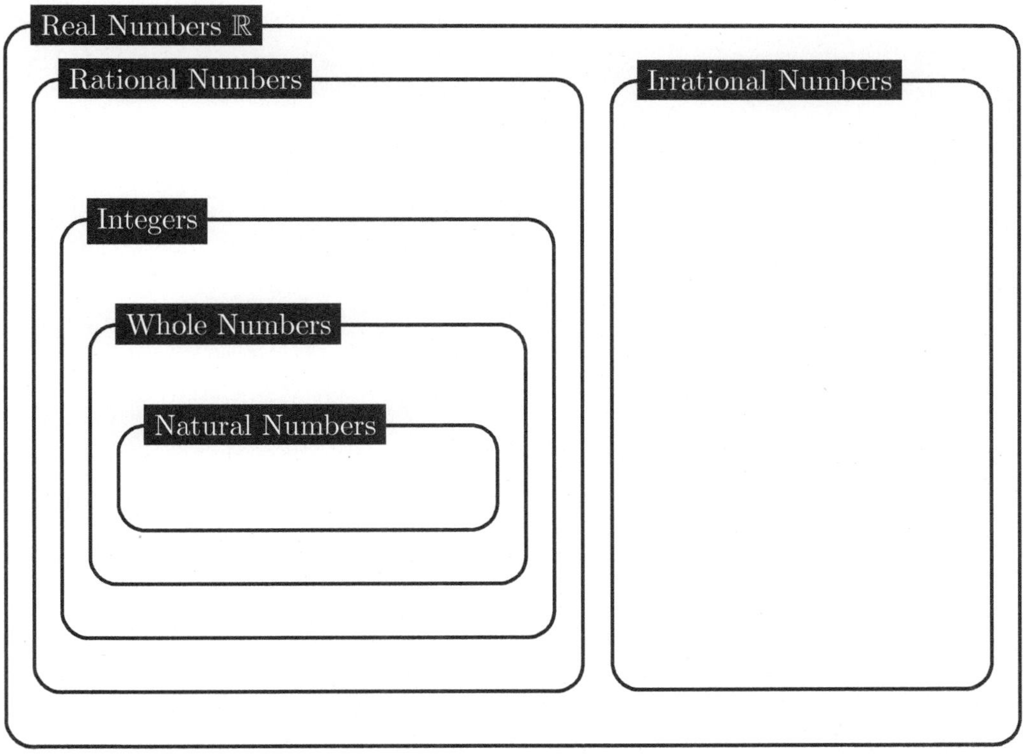

Every real number has a decimal representation. If the number is rational, then its corresponding decimal representation is *terminating* or *repeating*. For example,

- $\frac{3}{4} =$
- $\frac{9}{11} =$
- $\frac{9}{7} =$

If the number is irrational, then the decimal representation neither terminates nor repeats.

- $\pi =$
- $e =$
- $\sqrt{3} =$

II. Order and Interval/Inequality Notation

The real numbers can be represented by points on a line. The point corresponding to 0 is called the *origin*. A positive number x is represented by the point on the line x units to the right of the origin. A negative number $-x$ is represented by the point on the line x units to the left of the origin.

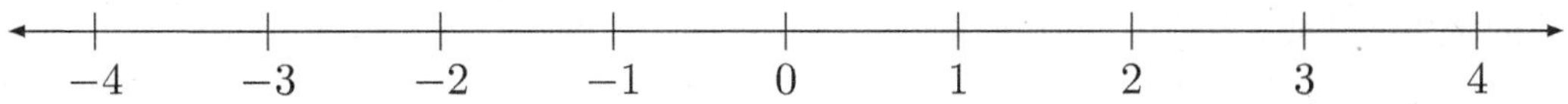

The ________________ of a number x, denoted $|x|$, is the *distance* from x to the origin.

$|x| =$ ____________ if $x \geq 0$ and $|x| =$ ____________ if $x < 0$.

Example 1 (Absolute value)

(a) $|3| =$

(b) $|-4| =$

The real numbers are *ordered*. Suppose that a and b are real numbers.

- If a lies to the left of b, we say that a is less than b and write ____________ .
- Equivalently, we can say that b is greater than a and write ____________ .

The statement $a \leq b$ means that

An *interval* is a particular subset of the real numbers that corresponds geometrically to a line segment. An interval may be *bounded* or *unbounded.*

Bounded Intervals

Let a and b be real numbers such that $a < b$. The following intervals on the real number line are *bounded.*

The ______________________ from a to b, denoted __________ , consists of all numbers between a and b (not including a and b).

The ______________________ from a to b, denoted __________ , consists of all numbers between **and including** a and b.

The numbers a and b are the *endpoints* of each interval. The parentheses ( ) indicate that endpoints are *excluded* from the interval; square brackets [ ] indicate that the endpoints are *included* in the interval.

The following table lists the different types of **bounded** intervals.

Interval Notation	Inequality Notation	Graph
(a, b)		
	$a \leq x \leq b$	
	$a \leq x < b$	
$(a, b]$		

Example 2 (Bounded intervals)

Complete the table.

Interval Notation	Inequality Notation	Graph

$[-2, 1]$		
	$3 < x < 4$	

Unbounded Intervals

A right endpoint of ∞ means that the interval continues forever in the positive direction; a left endpoint of $-\infty$ means that the interval continues forever in the negative direction. In either case, the interval is said to be *unbounded.*

The symbols ∞ (positive infinity) and $-\infty$ (negative infinity) are not real numbers. Instead, they are convenient symbols that describe the unboundedness of an interval. Since ∞ and $-\infty$ are not real numbers, they cannot be included in an interval of real numbers. So parentheses ( ) should always be used when an endpoint is ∞ or $-\infty$.

The following table lists the different types of **unbounded** intervals.

Interval Notation	Inequality Notation	Graph
(a, ∞)		
	$a \leq x$	
	$x < b$	
$(-\infty, b]$		
$(-\infty, \infty)$		

Example 3 (Unbounded intervals)

Complete the table.

Interval Notation	Inequality Notation	Graph
$[5, \infty)$		
	$-3 < x$	

III. Order of Operations

We use the phrase "**P**lease **E**xcuse **M**y **D**ear **A**unt **S**ally" to remember the *Order of Operations.*

Step 1: **P** - Parentheses, brackets, absolute value symbols, and radical symbols all act as *grouping* symbols that should be simplified from the inside out. The fraction bar also acts as a grouping symbol - simplify the numerator and denominator as separate groups.

Step 2: **E** - Exponents and radicals are simplified in order from ______________________ .

Step 3: **MD** - Multiplication and division are performed from ______________________ as they are encountered.
It is incorrect to first do all the multiplication and then do the division.

Step 4: **AS** - Addition and subtraction are also performed from ______________________ as they are encountered.
It is incorrect to first do all the addition and then do the subtraction.

Example 4 (Simplify using order of operations)

Simplify the expression.

$$3^2 - \sqrt{4}(-3) \div (8 \div 4)$$

Example 5 (Simplify using order of operations)

Simplify the expression.

$$\frac{1 - |-4 + 3|}{(1+4)^2}$$

Additional Notes

NAME: ______________________ EMAIL: ____________________ @vols.utk.edu

Lesson 2: The Basic Rules of Algebra

Learning Objectives:

I. Evaluate algebraic expressions and use correct vocabulary to identify the components of algebraic expressions.

II. Identify and apply the basic rules and properties of algebra.

III. Simplify fractional expressions.

IV. Apply divisibility rules.

V. Write the prime factorization of a number. Find the greatest common factor (GCF) and the least common multiple (LCM) of a set of numbers.

I. Algebraic Expressions

An *algebraic expression* is a collection of letters (called *variables*) and numbers (called *constants*) that are combined using the operations of addition, subtraction, multiplication, and division.

A *variable* is a symbol that represents an element of a set (its value can vary).

- The ______________ of an algebraic expression are those parts that are separated by addition.
- A term that contains variables is called a ______________ term.
- A term that consists of a number alone is called a ______________ term.
- The numerical factor of a variable term is the ______________ of the variable term.

Example 1 (Identify the components of an algebraic expression)

Consider the algebraic expression $5x^4 + 3x^2 - 2x + 7$.

(a) List the terms.

(b) What is the coefficient of the 3rd term?

(c) What is the variable part of the 1st term?

(d) What is the constant term?

To *evaluate* an expression, substitute given numerical values for each of the variables in the expression. When evaluating an algebraic expression, **remember to use the order of operations**.

Example 2 (Evaluate an algebraic expression)

Evaluate the algebraic expression $3x^2 - 5$ when $x = -1$.

Example 3 (Evaluate an algebraic expression)

Evaluate the algebraic expression $-x^2 + 2x$ when $x = -2$.

Example 4 (Evaluate algebraic expressions)

What is the difference between the expressions -4^2 and $(-4)^2$?

II. Basic Rules and Properties of Algebra

Addition and multiplication are the *primary* operations. Subtraction and division are defined in terms of these two operations. (So, when we say that terms are separated by addition, subtraction is included since subtraction means "add the opposite.")

Let a, b, and c be real numbers, variables, or algebraic expressions. The following properties are true.

- **Commutative Properties**
  - ☐ Addition:

    Example:
  - ☐ Multiplication:

    Example:

- **Associative Properties**

  ☐ Addition:

  Example:

  ☐ Multiplication:

  Example:

- **Distributive Property**

  ☐

  Example:

  ☐

  Example:

- **Identity Properties**

  ☐ Additive Identity: The number 0 is called the *additive identity* because any algebraic expression maintains its identity when the number 0 is added to it.

  Example:

  ☐ Multiplicative Identity: The number 1 is called the *multiplicative identity* because any algebraic expression maintains its identity when the number 1 is multiplied by it.

  Example:

- **Inverse Properties**

  ☐ Additive Inverse: The additive inverse of a is the expression $-a$ because the sum of an expression and its additive inverse equals the additive identity 0.

  Example:

  ☐ Multiplicative Inverse: The multiplicative inverse of a (assuming $a \neq 0$) is the expression $1/a$ because the product of an expression and its multiplicative inverse equals the multiplicative identity.

  Example:

Example 5 (Identify the property)

Which property is being illustrated?

(a) $(x + 5) + y = (5 + x) + y$

(b) $3(4x) = (3 \cdot 4)x$

III. Properties and Operations of Fractions

Let $a, b, c,$ and d be real numbers, variables, or algebraic expressions such that $b \neq 0$ and $d \neq 0$. The following properties are true.

1. Equivalent Fractions: $\frac{a}{b} = \frac{c}{d}$ if and only if ______________ .

   Example:

2. Rules of Signs: $-\frac{a}{b} =$ $\qquad$ $=$ $\qquad$ and $\frac{-a}{-b} =$

   Example:

3. Generate Equivalent Fractions: $\frac{a}{b} =$ $\qquad$, if $c \neq 0$.

   Example:

4. Add and Subtract: $\frac{a}{b} \pm \frac{c}{b} =$

   (Note: Must have common denominator! Use #3 to generate equivalent fractions first if the denominators are not the same.)

   Example:

5. Multiply Fractions: $\frac{a}{b} \cdot \frac{c}{d} =$

(Note: You do not need a common denominator to multiply fractions!)

Example:

6. Divide Fractions: $\frac{a}{b} \div \frac{c}{d} =$ (if $c \neq 0$)

Example:

Example 6 (Simplify fractional expressions)

Perform the operation and simplify the answer.
(Note: For fractions, "simplify" means reduce but do not write as a mixed number.)

(a) $-\frac{3}{7} + \frac{1}{4}$

(b) $\frac{3/4}{9}$

(c) $\left(\frac{1}{2}\right)^2 \div \frac{2}{3} + \frac{5}{3}$

IV. Divisibility Rules

A number a is *divisible* by a nonzero number b if there exists a real number c such that $a = bc$. (In other words, a is divisible by b if the remainder is zero when a is divided by b.)

We use divisibility rules to help determine whether a number is divisible by another without using a calculator. The divisibility rules say that a number is divisible by:

- 2 if the last digit is 0, 2, 4, 6, or 8.

  Example: 24 is divisible by 2 since the last digit is 2.

- 3 if sum of the digits is a multiple of 3.

  Example: 3654 is divisible by 3 since $3 + 6 + 5 + 4 = 18$ *is divisible by 3.*

- 4 if the last two digits are a multiple of 4.

  Example: 12364 is divisible by 4 since 64 is divisible by 4.

- 5 if the last digit is 0 or 5.

  Example: 35 is divisible by 5 since the last digit is 5.

- 6 if the number is divisible by BOTH 2 and 3.

  Example: 36 is divisible by 6 since it is divisible by 2 and 3.

- 8 if the last three digits are divisible by 8.

  Example: 12360 is divisible by 8 since 360 is divisible by 8.

- 9 if the sum of the digits is a multiple of 9.

  Example: 3654 is divisible by 9 since $3 + 6 + 5 + 4 = 18$ *is divisible by 9.*

- 10 if the last digit is 0.

  Example: 80 is divisible by 10 since the last digit is 0.

Example 7 (Determine divisibility)

Determine if the number 1347 is divisible by 2, 3, 4, 5, 6, 8, 9, and 10. Justify your answer.

V. Prime Factorization, GCF, and LCM

Factors are numbers that you multiply together to get another number. A *prime number* is a natural number (greater than 1) whose only factors are 1 and itself. *Prime Factoring* is the process of finding all of the prime factors of a number. We will do this using a *Factor Tree.*

Example 8 (Create factor trees and write prime factorization)

Create factor trees for 24, 64, and 54. Write each prime factorization using powers of primes.

The *greatest common factor (GCF)* of a collection of numbers is the largest number that is a factor of **all** numbers. The GCF is the product of the smallest power of the prime factors that all the numbers have in common. (The GCF is always less than or equal to all numbers.)

Example 9 (Find the GCF)

(a) Find the GCF of 6 and 15.

(b) Find the GCF of $40 = 2^3 \cdot 5$ and $60 = 2^2 \cdot 3 \cdot 5$.

(c) Find the GCF of $24 = 2^3 \cdot 3$ and $35 = 5 \cdot 7$.

The *least common multiple (LCM)* of a collection of numbers is the smallest number that is a multiple of **all** numbers. The LCM is the product of the highest power of each prime factor represented. The LCM is always greater than or equal to all numbers.)

Example 10 (Find the LCM)

(a) Find the LCM of 6 and 15.

(b) Find the LCM of $20 = 2^2 \cdot 5$ and $60 = 2^2 \cdot 3 \cdot 5$.

(c) Find the LCM of $36 = 2^2 \cdot 3^2$ and $120 = 2^3 \cdot 3 \cdot 5$.

Additional Notes

NAME: ______________________ EMAIL: __________________ @vols.utk.edu

Lesson 3: Integer Exponents

Learning Objectives:

I. Identify the properties of integer exponents.

II. Simplify expressions containing integer exponents.

I. Properties of Integer Exponents

If a is any real number and n is a positive integer, then the nth *power* of a is

The number a is called the ______________, and n is called the ______________.

The following are rules for working with exponents. The bases a and b are real numbers, and the exponents m and n are integers.

1. $a^m \cdot a^n =$

Ex:

2. $\dfrac{a^m}{a^n} =$

Ex:

3. $(ab)^n =$

Ex:

4. $\left(\dfrac{a}{b}\right)^n =$

Ex:

5. $(a^m)^n =$

Ex:

6. $a^0 =$

Ex:

7. $a^{-n} =$

Ex:

8. $\left(\dfrac{a}{b}\right)^{-n} =$

Ex:

II. Simplify Expressions

Example 1 (Simplify expressions with integer exponents)

Use the properties of exponents to simplify the following expressions, leaving only positive exponents.

(a) $(-5z)^3$

(b) $(8x^4)(2x^3)$

(c) $(-z)^3(3z^4)$

(d) $\left(\frac{4}{y}\right)^3\left(\frac{3}{y}\right)^4$

(e) $\frac{15(x+3)^3}{9(x+3)^2}$

(f) $(x+5)^0$, $x \neq -5$

(g) $\left(\frac{1}{2}\right)^{-3}$

(h) $5^{-1} - 2^{-1}$

(i) $4x^{-2}$

(j) $\frac{9a^2b^{-2}}{3a^{-1}b}$

(k) $\left(\frac{x}{10}\right)^{-1}$

(l) $\left(\frac{x^{-3}y^4}{5}\right)^{-3}$

Additional Notes

NAME: ______________________ EMAIL: ____________________ @vols.utk.edu

Lesson 4: Radicals and Rational Exponents

Learning Objectives:

I. Identify the properties of radical expressions and simplify an expression containing radicals.

II. Convert between radical and rational exponent form.

III. Identify the properties of rational exponents and simplify an expression containing rational exponents.

IV. Evaluate an expression containing rational exponents.

I. Properties of Radical Expressions

The symbol $\sqrt{}$ means "the positive square root of." Thus, $\sqrt{a} = b$ means that

The symbol $\sqrt{a}$ only makes sense when $a \geq 0$ since $a = b^2 \geq 0$.

For example, $\sqrt{9} =$

Square roots are special cases of nth roots. If n is any positive integer, then the *principal* n^{th} *root* of a is defined as follows:

The positive integer n is the ______________ and the number a is the ______________ .

When n is even, we must have that

When n is odd, there are no restrictions on a or b.

For example,

- $\sqrt[4]{16}$

- $\sqrt[3]{-27}$

- $\sqrt[4]{-16}$

The following are rules for working with radicals. Let a and b be positive real numbers, and let m and n be positive integers.

1. $\sqrt[n]{a^m} = (\sqrt[n]{a})^m$

Ex:

2. $\sqrt[n]{ab} = \sqrt[n]{a} \cdot \sqrt[n]{b}$

Ex:

3. $\sqrt[n]{\frac{a}{b}} = \frac{\sqrt[n]{a}}{\sqrt[n]{b}}$

Ex:

4. $\sqrt[n]{a^n} =$

Ex:

Example 1 (Simplify radical expressions)

Simplify each expression. *Note: For square roots, "simplify" means the final answer should have a single square root such that the radicand has no perfect square factors.*

(a) $\sqrt{72}$

(b) $\sqrt{\frac{5}{9}}$

(c) $(\sqrt{11})^2$

(d) $\sqrt{18}$

(e) $\sqrt{27} \cdot \sqrt{3}$

(f) $-\sqrt{\frac{16}{81}}$

II. Convert Between Radical and Rational Exponent Form

We use the following properties to convert between radical and rational exponent form.

1. $\sqrt[n]{a^m} =$ Ex:	**2.** $\sqrt[n]{a} =$ Ex:

Note that Property #2 is a special case of Property #1 (when there is no exponent inside the radical).

Example 2 (Convert between radical and rational exponent form)

Complete the table.

Radical Form	Rational Exponent Form
$\sqrt[5]{2^4}$	
	$(-81)^{1/3}$
$-\sqrt[3]{81}$	
	$-3^{1/2}$
$\sqrt[4]{x^3}$	

III. Properties of Rational Exponents

The exponent rules for integer exponents (see Lesson 3) also apply to rational exponents.

Example 3 (Simplify rational exponent expressions)

Simplify each expression. Leave answers with only positive exponents.

(a) $(x^3y^2)^{3/2}$

(b) $\left(\dfrac{x^{1/3}}{x^{3/5}}\right)^{-2}$

(c) $\dfrac{x^{2/3}\,y^{3/4}}{x^{1/6}\,y^2}$

IV. Evaluate Expressions with Rational Exponents

When evaluating expressions that involve rational expressions without a calculator, it is helpful to convert to radical form and/or use the properties of rational exponents to simplify before evaluating.

Example 4 (Evaluate rational exponent expressions)

Simplify each expression.

(a) $8^{2/3}$

(b) $-81^{3/4}$

(c) $\dfrac{9^{2/3}}{9^{1/6}}$

(d) $125^{-2/3}$

Additional Notes

NAME: ______________________ EMAIL: ____________________ @vols.utk.edu

Lesson 5: Polynomial Expressions and Special Products

Learning Objectives:

I. Use correct vocabulary to identify the components and types of polynomial expressions.

II. Add and subtract polynomial expressions.

III. Multiply polynomial expressions and recognize basic special products.

I. Polynomial Expressions

A *polynomial expression* is an expression of the form

where $a_0, a_1, \ldots, a_n$ are real numbers, and n is a non-negative integer.

If $a_n \neq 0$, then:

- n is the ______________ of the polynomial expression,
- and a_n is the ____________________ .

Polynomial expressions with:

- one term are called ____________________ .
- two terms are called ____________________ .
- three terms are called ____________________ .

Recall: The *terms* of an algebraic expression are those parts that are separated by addition (or subtraction).

Example 1 (Components of polynomial expressions)

Identify the degree, leading coefficient, and constant term of the polynomial expression.

$$14x^5 - 3x^4 + 3x^2 - 7x + 4$$

Degree = _______ Leading Coefficient = _______ Constant Term = _______

A polynomial expression is in *standard form* if each coefficient is in simplest form and the terms are arranged from ______________ power to ______________ power.

Example 2 (Standard form and components of polynomial expressions)

Consider the polynomial expression $7x - \frac{3}{9}x^3 + \pi x^4$.

(a) Write the polynomial expression in standard form.

(b) Identify the degree, leading coefficient, and constant term of the polynomial expression.

Degree = ________ Leading Coefficient = ________ Constant Term = ________

II. Add and Subtract Polynomial Expressions

To add or subtract polynomial expressions, combine like terms.

Example 3 (Add and subtract polynomial expressions)

Perform the operation and write the result in standard form.

(a) $(3x^3 - 4x^2 + 5x - 1) + (5x^3 + 3x^2 - 3x + 7)$

(b) $(3x^3 - 4x^2 + 5x - 1) - (5x^3 + 3x^2 - 3x + 7)$

III. Multiply Polynomial Expressions and Special Products

To multiply polynomial expressions, use the distributive property.

Example 4 (Multiply polynomial expressions)

Perform the operation and write the result in standard form.

(a) $3x^2(x^4 + 5x - 7)$

(b) $(2x + 1)(3x - 4)$

(c) $(3x - 1)(2x^2 + 3x - 1)$

Special Products: Let u and v be real numbers, variables, or algebraic expressions.

Sum and Difference of Same Terms:

$(u + v)(u - v) =$

Square of a Binomial:

$(u + v)^2 =$
$=$

$(u - v)^2 =$
$=$

Example 5 (Special products)

Perform the operation and write the result in standard form.

(a) $(3x+4)(3x-4)$

(b) $(x-\frac{1}{2})(x+\frac{1}{2})$

(c) $(x-2)^2$

(d) $(2x+1)^2$

Additional Notes

NAME: ______________________________ EMAIL: ________________________ @vols.utk.edu

Lesson 6: Factoring

Learning Objectives:

I. Factor polynomial expressions by extracting the greatest common factor.

II. Factor polynomial expressions containing a difference of two squares.

III. Factor polynomial expressions by grouping.

IV. Factor trinomial expressions with binomial factors.

I. Common Factors

Factoring is the process of writing a polynomial as a product. It is an important tool for solving equations and reducing fractional expressions.

Removing a *common factor* is the simplest type of factoring and involves using the distributive property (in the reverse direction):

$$ab + ac =$$

Example 1 (Factor by extracting common factors)

Factor completely.

(a) $9x^2 - 6x =$

(b) $6x^4 + 2x^3 =$

(c) $x^2(x-2) + 3(x-2) =$

II. Difference of Two Squares

Let u and v be real numbers, variables, or algebraic expressions.

$u^2 - v^2 =$

Example 2 (Factor a diference of two squares)

Factor completely.

(a) $4x^2 - 121$

(b) $64x^2 - 1$

(c) $5x^2 - 20$

(d) $3x^4 - 27x^2$

(e) $4 - x^2$

(f) $32x - 2x^3$

III. Factor by Grouping

Factoring by grouping involves removing a common factor from a group of terms.

Example 3 (Factor by grouping)

Factor completely.

(a) $x^6 + 3x^4 + x^2 + 3$

(b) $3x^3 - 6x^2 - 2x + 4$

(c) $x^3 + 4x^2 - x - 4$

IV. Factor Trinomials with Binomial Factors

The **ac method** is used to factor trinomials in the form $ax^2 + bx + c$. Here are the steps:

1. Find the product of the leading coefficient and the constant term (ac).

   Example: Find ac for the trinomial $3x^2 + x - 4$.

2. List the factors of ac and find the pair whose sum is b.

   Example: List the factors of ac and find the pair whose sum is 1.

3. Rewrite the bx term as two terms using the pair of coefficients identified in Step 2.

   Example: Rewrite the trinomial by replacing the term x with two terms.

4. Factor by grouping.

   Example: Factor the rewritten trinomial by grouping.

Example 4 (Factor trinomials with binomial factors)

Factor completely.

(a) $8x^2 - 10x - 3$

(b) $10x^2 + 19x + 6$

(c) $-6y^2 - 3y + 3$

(d) $w^2 - 10w + 21$

Additional Notes

NAME: ______________________ EMAIL: ______________________ @vols.utk.edu

Lesson 7: Rational Expressions

Learning Objectives:

I. Determine whether an expression is a rational expression.

II. Determine the domain of a polynomial or rational expression.

III. Simplify a rational expression by dividing out common factors.

IV. Multiply and divide rational expressions.

V. Rewrite a quotient so it has no variables in the denominator.

I. Rational Expressions

A *rational expression* is the ______________________ of two polynomials.

Example 1 (Rational expressions)

Determine whether each expression is a rational expression or not. Explain.

(a) $\dfrac{3x^2 + 7x + 5}{x - 4}$

(b) $\dfrac{\sqrt{x}}{x^2 + 4}$

II. Domain

The ______________________ of an algebraic expression is the set of all real numbers for which the expression is defined as a real number when evaluated.

- The domain of a **polynomial expression** is the set of ______________________ ,

  denoted ______________________ using interval notation.

- The domain of a **rational expression** is the set of all real numbers except the values that

  ______________________ .

Domain restrictions are the real numbers that are **not** in the domain of an algebraic expression.

"Domain" and "domain restrictions" are opposites. You can think of the domain being all the real numbers you can plug into the expression; the domain restrictions are the real numbers you cannot plug into the expression.

Example 2 (Domain of polynomial and rational expressions)

Determine the domain of each expression. Give the answer as indicated.

(a) $3x^2 - 7x + 4$ (in interval notation)

(b) $\dfrac{x}{x+2}$ (in words)

(c) $\dfrac{3x+14}{-x+3}$ (in interval notation)

III. Simplifying Rational Expressions

To simplify a rational expression:

- ____________________ the numerator and denominator.
- List the domain restrictions. Since two algebraic expressions must have the same domains to be equivalent, **the domain restrictions come from the original expression**.
- Simplify by ____________________ common factors.

Remember, factors are attached to each other by ____________________ .

Terms are attached by ____________________ .

You cannot cancel <u>terms</u> that are common to both the numerator and denominator!!

Example 3 (Simplify rational expressions)

Simplify the rational expression. List all domain restrictions.

(a) $\dfrac{x^2 - 4x + 4}{x^2 - 4}$

(b) $\dfrac{2x^2 - 5x - 3}{x^3 - 3x^2}$

IV. Multiplying and Dividing Rational Expressions

Multiply and divide rational expressions using the rules of fractions from Lesson 2.

Example 4 (Multiply and divide rational expressions)

Perform the operation and simplify. List all domain restrictions.

(a) $\dfrac{3y - 6}{5y + 35} \cdot \dfrac{2y + 7}{2 - y}$

(b) $\dfrac{x^2 - 5x + 6}{3x - 6} \div \dfrac{x + 6}{3x - 6}$

V. Rewriting Quotients

It is often useful to rewrite a quotient so that it has no variables in the denominator. To do so, you may need to use negative exponents. *Note: This is a very important skill in Math 125.*

Example 5 (Rewrite quotients)

Rewrite each expression so that there are no variables in the denominator.

(a) $\dfrac{5x^3 + 3x^2 - 4x + 7}{x}$; $x \neq 0$

(b) $\dfrac{5x^3 + 3x^2 - 9}{\sqrt{x}}$; $x \neq 0$

(c) $\dfrac{3}{(x^2 + 5)^4}$

(d) $\dfrac{3}{x^2} + \dfrac{4}{\sqrt{x-1}} - \dfrac{1}{2x}$; $x \neq 0, 1$

Additional Notes

Unit 2

Equations and Inequalities

In this unit, we build on the knowledge gained in Unit 1. We focus on developing a strong understanding of equations and inequalities as essential tools for analyzing mathematical relationships. We will solve various types of equations (linear, quadratic, and polynomial) and linear inequalities in one variable.

NAME: ______________________ EMAIL: __________________ @vols.utk.edu

Lesson 8: Linear Equations

Learning Objectives:

I. Classify an equation as an identity, a conditional, or a contradiction equation.

II. Solve a linear equation in one variable.

III. Solve a literal equation for an indicated variable.

I. Identity vs. Conditional Equations

An __________________ is a statement that two algebraic expressions are *equal.* Up to this point, we have dealt with algebraic expressions (no equal sign).

Example 1 (Expression vs. equation)

Give one example of an algebraic expression and one example of an equation.

To *solve* an equation means to find all values of the variable for which the equation is *true.*

- An __________________ equation is an equation that is true for *every* real number in the domain of the variable. (The equation has infinitely many solutions.)
- A __________________ equation is an equation that is true for just *some* (or even none) of the real numbers in the domain of the variable. (The equation has finitely many solutions.)
- A __________________ equation is a equation that is true for *none* of the real numbers in the domain of the variable. It is a special type of conditional equation. (The equation has no solution.)

To determine whether an equation is an identity equation or a conditional equation, *simplify* each side of the equation as much as possible. If both sides are identical, then it is an identity equation. If not, it is a conditional equation.

Example 2 (Identity vs. conditional equation)

Determine whether each equation is an identity or conditional equation. Show work to support your answer.

(a) $3x + 2(x-1) = 5x - (x+2)$

(b) $5x - (2x-9) = 3(x+3)$

II. Solve Linear Equations

A *linear equation* in one variable has a polynomial of degree no greater than one on each side of the equation. To solve linear equations, we produce equivalent equations - equations with the same solutions - until we determine the solution. There are four ways to produce equivalent equations:

- Replace any expression with an equivalent one using the rules of algebra from Unit 1.
- Add (or subtract) the same quantity to (or from) both sides.
- Multiply (or divide) both sides of the equation by the same nonzero quantity.
- Interchange the two sides of the equation.

To solve a linear equation, we usually begin by ____________________ each side of the equation as much as possible. If the variable appears on both sides of the equation, add (or subtract) quantities from both sides so that all of the variable terms are on one side. Then isolate the variable by removing other things on that side working your way from outside in - sort of like reverse order of operations.

Note: You can check if an answer is correct by plugging the answer into the original equation. If this results in a true statement, then the answer is correct.

Example 3 (Solve linear equations)

Solve.

(a) $7 - 2x = 15$

(b) $2(x+5)-7=3(x-2)$

(c) $-3(2x+3)=8-5x$

(d) $\frac{5x}{4}+\frac{1}{2}=x-\frac{1}{2}$

III. Literal Equations

A *literal equation* is an equation in which variables represent known values. Literal equations are used to give formulas such as distance, speed, area, volume, force, time, temperature, etc. Sometimes you are asked to solve a literal equation for a specified variable. Use the same rules you use for solving a linear equation to isolate the variable specified.

Example 4 (Solve literal equations)

Solve for the specified variable.

(a) $T = D + pm$ for p

(b) $A = \frac{1}{2}h(a + b)$ for a

(c) $IR + Ir = E$ for I

(d) $-5(3x - 6b) + 12 = 8 + 3ax$ for x

(e) $A = P(1 + rt)$ for t

Additional Notes

NAME: ______________________ EMAIL: ____________________ @vols.utk.edu

Lesson 9: Quadratic Equations and the Quadratic Formula

Learning Objectives:

I. Solve a quadratic equation by factoring.

II. Solve a quadratic equation by extracting square roots.

III. Solve a quadratic equation using the Quadratic Formula.

IV. Calculate the discriminant to determine the number of real solutions of a quadratic equation.

I. Solve by Factoring

A *quadratic equation* is any equation that can be written in the form ____________________ , where a, b, and c are real numbers with $a \neq 0$. This form is called the *general form.*

To solve an equation by *factoring*, you must use the **Zero-Factor Property**, which says that:

Note: The equation must be written so that one side is equal to 0 before applying the Zero-Factor Property. The factoring method only works if the quadratic expression will factor.

Example 1 (Solve by factoring)

Solve by factoring.

(a) $x^2 + 3x + 2 = 0$

(b) $3x^2 - 12x = 0$

(c) $-x^2 + 5x = 6$

II. Solve by Extracting Square Roots

Extracting Square Roots is a shortcut for solving equations of the form $u^2 = d$, where u is an algebraic expression and d is a non-negative constant.

To use this method, isolate the squared part. Then take the square root of both sides. You *must* put $\pm$ in front of the solution side or you will not have all the solutions.

Why? Think about the equation $x^2 = 16$. Is $x = 4$ the only solution to this equation?

Example 2 (Solve by extracting square roots)

Solve by extracting square roots.

(a) $3x^2 = 36$

(b) $(x+2)^2 = 27$

(c) $12x^2 = 300$

III. Solve Using the Quadratic Formula

If you are unable to solve by factoring or extracting square roots, you must use the *Quadratic Formula.* The Quadratic Formula works for every quadratic equation, but should only be used as a last resort.

The Quadratic Formula states that the solution(s) of the quadratic equation $ax^2 + bx + c = 0$ are given by:

Example 3 (Solve using the Quadratic Formula)

Solve using the Quadratic Formula.

(a) $x^2 + 8x - 4 = 0$

(b) $4x^2 - 4x = 7$

IV. The Discriminant

The *discriminant,* ____________________ , is an expression that can be used to determine the number of real solutions for a quadratic equation. The variables a, b, and c are the coefficients in the general form of the equation: $ax^2 + bx + c = 0$.

$b^2 - 4ac > 0 \Rightarrow$	two real solutions
$b^2 - 4ac = 0 \Rightarrow$	one real solution
$b^2 - 4ac < 0 \Rightarrow$	no real solutions

Tip: Calculating the discriminant can help you determine the best way to solve the quadratic equation. If the discriminant is a perfect square, then the quadratic expression will factor with integer coefficients.

Given a quadratic equation, how do we determine the best way to solve it?

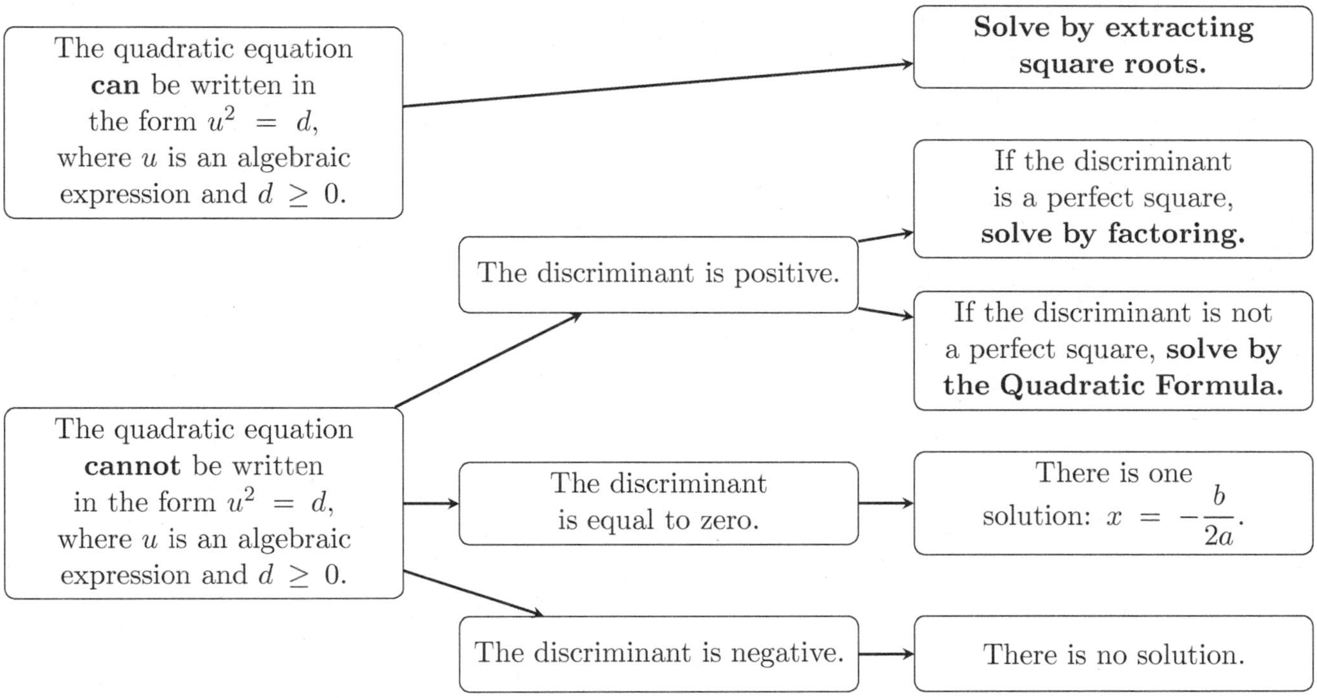

Example 4 (Number of solutions and method of solving)

Use the discriminant to determine the number of real solutions of each quadratic equation. What is the best way to solve each equation?

(a) $3x^2 + 4x - 1 = 0$

(b) $x^2 + 3x + 2 = 0$

Additional Notes

NAME: ____________________ EMAIL: ____________________ @vols.utk.edu

Lesson 10: Other Types of Equations

Learning Objectives:

I. Solve a polynomial equation by factoring.

II. Solve an equation containing rational expressions.

I. Solve Polynomial Equations by Factoring

To solve polynomial equations by factoring, use the **Zero-Factor Property**. Remember the equation must be written so that one side is equal to 0 before applying this property.

Example 1 (Solve polynomial equations by factoring)

Solve.

(a) $x^3 = 4x$

(b) $x^3 - 7x^2 - 4x + 28 = 0$

(c) $x^3 + 5x = 0$

Example 2 (Write an equation with given solutions)

Write an equation in standard form that has the given solutions: -3, 5, and 0

II. Solving Equations Containing Rational Expressions

If an equation contains rational expressions, begin by ______________________________ .
The least common denominator (LCD) is the least common multiple of the denominators.

To determine the LCD, first factor all of the ______________________ . The LCD contains each different factor you see raised to the highest power you see.

Note: Since the domain of the expression is not necessarily all real numbers, you must remember to check for *extraneous solutions* (i.e., check to make sure that the solution(s) you find are actually in the domain of the original equation).

Example 3 (Solve equations containing rational expressions)

Solve. $\dfrac{6}{x^2-4x+3} = \dfrac{3}{x-3} + \dfrac{4}{x-1}$

Example 4 (Solve equations containing rational expressions)

Solve. $\dfrac{1}{x} = \dfrac{4}{x-1} + 1$

Example 5 (Solve equations containing rational expressions)

Solve. $\frac{4}{x} - \frac{5}{3} = \frac{x}{6}$

Example 6 (Solve equations containing rational expressions)

Solve. $\frac{4}{x+1} - \frac{3}{x+2} = 1$

Example 7 (Solve equations containing rational expressions)

Solve. $\dfrac{x+2}{x-2} - \dfrac{x}{3} = 0$

Additional Notes

NAME: ______________________________ EMAIL: ________________________ @vols.utk.edu

Lesson 11: Modeling with Linear Equations

Learning Objectives:

I. Set up and solve an application problem involving percentages.

II. Set up and solve an application problem involving common formulas.

III. Set up and solve an application problem involving the cost, revenue, and profit equations.

IV. Set up and solve a general linear mathematical modeling problem.

V. Translate a word problem into a system of linear equations.

I. Application Problems Involving Percentages

In this section, we will use algebra to solve real-life problems. *Mathematical modeling* is the process of constructing one or more equations that represent real-life problems.

Some words have specific mathematical meaning:

- ________________________ indicate equality
- ________________________ indicates multiplication
- Others are listed in the textbook.

We solve problems involving percentages by translating the key words into an equation we can solve. When doing calculations involving percentages, you should convert to decimal or fraction form.

Example 1 (Convert percentages to decimals and fractions)

Convert each percentage into a decimal and a fraction.

(a) 34%

(b) 5%

(c) 0.04%

Example 2 (Solve percent equations)

Solve the percent equation.

(a) 150 is 5% of what number?

(b) What is 34% of 300?

II. Application Problems Involving Common Formulas

Sometimes, solving a word problem involves using a commonly known formula. The common formulas you need to memorize for this course are:

- Area of a Rectangle: ____________________
- Perimeter of a Rectangle: ____________________
- Distance: ____________________
- Simple Interest: ____________________
- Cost, Revenue, and Profit (covered in the next objective)

Example 3 (Represent given information using common formulas)

Use a common formula to write an equation to represent the given information.

(a) The distance d, traveled in t hours by a car traveling 60 miles per hour.

(b) The area, A, of a rectangle with length 3 cm and width x.

(c) The perimeter, P, of a rectangle with width x and a length that is twice the width.

(d) The interest earned if x dollars are invested for one year at a rate of 12% simple interest per year.

III. Application Problems Involving the Cost, Revenue, and Profit Equations

The *cost* equation gives the total cost of producing a quantity/number of units of some product. Costs can be separated into two parts: the *fixed* costs, which are incurred even if nothing is produced, and the *variable* cost, which varies depending on how many units are produced.

Cost = (fixed cost) + (variable cost)
Variable Cost = (cost per unit) · (number of units)

The *revenue* equation gives the total revenue received by the company from selling a quantity of some product.

Revenue = (price per unit) · (number of units)

The *profit* equation gives the total profit received from producing and selling a quantity of some product.

Profit = (Revenue) − (Cost)

Note: Since the number of items varies, a variable is used to represent the quantity or number of items/units.

Example 4 (Cost, revenue, and profit equations)

Your friend Trevor is starting a new business, Trevor's Traveling Tacos. He plans to sell tacos out of a food truck that he recently purchased. The food truck and appliances required to begin production cost a total of $30,000. The costs of labor and raw materials are $5 per taco tray. Trevor plans to sell taco trays for $7 each.

(a) Write the total cost equation, C, for producing x taco trays.

(b) Write the revenue equation, R, for selling x taco trays.

(c) Write and simplify the profit equation, P, for producing and selling x taco trays.

Note: Always simplify the profit equation (subtract the polynomials). If you are asked to write an equation, do not leave off the $C =$, $R =$, or $P =$ in the front. The equal sign is what makes it an equation rather than an expression.

IV. Mathematical Modeling

Solving more complicated word problems often involves assigning labels or defining variables. In the examples below, we will use a verbal model and labels to help us write an equation to solve.

Example 5 (Set up and solve mathematical modeling problems)

(a) A shirt that normally sells for \$40 is marked down \$8. What percent of the original price is the markdown?

Verbal Model:

Labels: Equation:

Solution:

(b) A shirt that normally sells for \$40 is marked down to \$8. What percent of the original price is the markdown?

How is this problem different from the previous problem?

Verbal Model:

Labels: Equation:

Solution:

(c) If you average 70 miles per hour, how long will it take you to drive 210 miles?

Verbal Model:

Labels: Equation:

Solution:

(d) You are taking a class that has four tests. The first three tests are 100 points each and the fourth test is 200 points. To get a B in the course, you must have an average of at least 80% on the four tests. Your scores on the first three tests were 78, 82, and 72. What must you score on the fourth test to get a B in the course?

Verbal Model:

Labels:

Equation:

Solution:

(e) Sarah's weekly paycheck is 20% less than her coworker Linda's paycheck. The two paychecks total $1800. Find the amount of each check.

Verbal Model:

Labels:

Equation:

Solution:

V. Translate Word Problems into Systems of Equations

In Math 123, you'll need to be able to translate word problems into a system of equations. A system of equations involves more than one variable and more than one equation. (We will not learn to solve systems of equations in this course.) The emphasis now will be on **defining the variables** needed and **writing the equations** to represent the problem. To define the variables, read the problem and focus on:

You must be **specific** when defining variables. Include units when possible.

Example 6 (Translate word problems into systems of equations)

Define the variables and write the problem as a system of equations. **Do not solve.**

(a) Brenda's school is selling tickets to a spring musical. On the first day of ticket sales, the school sold 3 senior citizen tickets and 9 child tickets for a total of \$75. The school took in \$67 on the second day by selling 8 senior citizen tickets and 5 child tickets. What is the price each for senior citizen tickets and for child tickets?

Define Variables: System of Equations:

(b) A total of \$15,000 is invested in two corporate bonds, Bond I and Bond II. Bond I earns 7.5% simple interest. Bond II earns 10% simple interest. The investor wants an annual interest income of \$1400 from the investments. How much should be invested in each type of bond?

Define Variables: System of Equations:

Additional Notes

NAME: ______________________ EMAIL: __________________ @vols.utk.edu

Lesson 12: Linear Inequalities in One Variable

Learning Objectives:

I. Represent an interval of numbers graphically and using specified notation; determine whether intervals are bounded or unbounded.

II. Determine whether a value is a solution to an inequality.

III. Solve single and double inequalities; describe solutions using a graph, inequality notation, and interval notation.

I. Interval vs. Inequality Notation

Recall (from Lesson 1) that interval notation and inequality notation are both ways to express a set of numbers from the real number line. An interval is *bounded* if it has real numbers for both endpoints. Otherwise, the interval is *unbounded.*

Example 1 (Represent intervals)

Complete the table. Represent each interval of numbers graphically and using specified notation. Determine whether the interval is bounded or unbounded.

Interval Notation	Inequality Notation	Graph	Bounded?
$[3, 7)$			
$(-\infty, 5)$			
	$4 \leq x$		
	$-2 < x < 1$		

II. Determine Whether a Value is a Solution to an Inequality

- A ______________________ of an inequality is a value for the variable that makes the inequality a true statement.
- The ______________________ of an inequality is the set of *all* solutions of the inequality.
- The ______________________ of an inequality is the set of all points on the real number line that correspond to a solution of the inequality.

Example 2 (Determine whether a value is a solution to an inequality)

Determine whether each value of x is a solution of the inequality. $0 < \frac{x-2}{4} < 3$

(a) $x = 10$

(b) $x = 0$

(c) $x = \frac{7}{2}$

III. Solve Linear Inequalities (Single & Double)

Solving linear inequalities is similar to solving linear equations. You can add or subtract the same number from both sides. You can multiply or divide both sides by the same non-zero number. BUT, if you multiply or divide by a *negative* number, you must reverse the inequality symbol.

Why? Let's consider the inequality $-4 > -8$, which is ______________ .

If we multiply both sides by -2 and forget to reverse the symbol, we get:

$(-2)(-4) > (-2)(-8) \Rightarrow 8 > 16$, which is ______________ .

In order to keep the inequality true, we must ______________ the inequality symbol to get $8 < 16$.

Example 3 (Solve single linear inequalities)

Solve. Give your answer as indicated.

(a) $3x + 1 \leq 7$ (Interval Notation)

(b) $4 - 3x \leq 10$ (Inequality Notation)

(c) $3 + 2x > 1$ (Graph)

(d) $2 - 5x > 6x + 8$ (Interval Notation)

Example 4 (Solve double linear inequalities)

Solve. Give your answer as indicated.

(a) $3 \leq 2x - 1 < 7$ (Inequality Notation)

(b) $-4 < \dfrac{2x-3}{3} < 4$ (Interval Notation)

Example 5 (Solve an application inequality problem)

Solve the word problem involving an inequality. The cost of traveling by taxi in Las Vegas is \$3.20 for the initial charge and \$2 for each mile driven. The cost in Philadelphia is \$2.30 for the initial charge and \$2.10 for each mile driven. How many miles must a passenger travel by taxi in Philadelphia for the cost to be more than traveling the same distance by taxi in Las Vegas?

Additional Notes

Unit 3

Functions and Graphs

In this unit, we broaden our scope of the knowledge gained in Units 1 and 2 to include the connection between functions and their graphs. We begin with topics such as interpreting and sketching graphs of functions and move on to apply transformations and examine piecewise functions. Lastly, we identify key features of polynomial functions, such as turning points and end behavior. By the end of this unit, students will be equipped to analyze and graph various functions.

NAME: ______________________ EMAIL: ______________________ @vols.utk.edu

Lesson 13: Functions

Learning Objectives:

I. Recognize the definition of a function and determine whether a relationship represents a function.

II. Identify the domain of a function algebraically.

III. Recognize and use correct function notation to evaluate a function algebraically.

IV. Find the zeros of a function algebraically.

I. Definition of a Function

The idea of a function is based on the fact that one quantity often depends on another. For example, one's height depends on their age, the air temperature depends on the date, and the cost of mailing a package depends on its weight. We use the term *function* to describe this dependence of one quantity on another.

What are some other examples of functions?

-

-

A ______________________ is correspondence in which each input x is assigned to one and only one output, called y. *(Not more than one, not less than one, exactly one.)*

Each function has a name, like f or g, but the letter name of the function is **not** a variable. If the function is named f, then the symbol $f(x)$ is another way of denoting the output corresponding to the input x.

It is helpful to think of a function as a machine. The *input* x enters the machine and the machine produces an *output* $y = f(x)$ according to the rule of the function.

The symbol $f(x)$ (read "f of x") is called the ______________________ .

The **domain** is the set of all inputs into the function, and the **range** is the set of all outputs.

An arrow diagram can be used to show the correspondence between each input and output.
Tip: Do not list repeated elements more than once when drawing an arrow diagram.

Example 1 (Determine whether a relationship represents a function)

Consider $S = \{(1, 3), (-3, 4), (6, 3)\}$.
Draw an arrow diagram for S.

Domain **Range**

Is S a function? Why or why not?

Consider $T = \{(1, 2), (2, 3), (3, 4), (2, 5)\}$.
Draw an arrow diagram for T.

Domain **Range**

Is T a function? Why or why not?

An equation in two variables x and y represents y as a function of x if the equation can be uniquely solved for y.

Example 2 (Determine whether an equation represents a function)

Do the following equations represent y as a function of x?

(a) $5x^2 + y - 7 = 0$

(b) $x + y^2 - 4 = 0$

II. Domain of a Function

Unless otherwise specified, the **domain** of a function is the set of all x-values such that the corresponding y-value is a real number. (Think of domain the same way you did in Unit 1, only now we use function notation.)

- Polynomials: the domain is ______________ .

- Fractions: values for x that make the denominator equal ________ are not in the domain.

- Radicals: For radicals with **odd indices**, the domain is ______________ (assuming the radicand is a polynomial).

  For radicals with **even indices**, the domain is the set of real numbers such that the

  ______________________ is ______________________ .

Example 3 (Identify domains)

Find the domain of each function. Give the answer as indicated.

(a) $f(x) = 5x^4 + 7x^3 - 2x^2 + 1$ (in words)

(b) $g(x) = \dfrac{3x}{x+2}$ (in interval notation)

(c) $h(x) = \sqrt{x-7}$ (in interval notation)

(d) $j(x) = \sqrt[4]{2-3x}$ (in inequality notation)

(e) $k(x) = \dfrac{\sqrt{x}}{x^2-4}$ (in interval notation)

III. Evaluating Functions

To **evaluate** a function f means to find the value of $f(x)$ for a given value of x. To do so, substitute the given value in place of x and simplify.

Example 4 (Describe and evaluate a function)

A function f is defined by the formula $f(x) = x^2 + 2$.

(a) Express in words how f acts on the input x to produce the output $f(x)$.

(b) Evaluate $f(2)$, $f(-1)$, and $f(\sqrt{3})$.

(c) Find the domain and range of f.

Example 5 (Evaluate a function)

Find the following, given $f(x) = x^2 - 3x$.

(a) $f(3) =$

(b) $f(0) =$

(c) $f(a) =$

(d) $f(x+2) =$

Example 6 (Evaluate a function)

Find the following, given $g(x) = -x + 4$.

(a) $g(1) =$

(b) $g(-2) =$

(c) $g(t) =$

(d) $g(x - 1) =$

Example 7 (Determine the values that make two functions equal)

If $f(x) = 2x^2$ and $g(x) = x^4 - 2x^2$, find all real values of x such that $f(x) = g(x)$.

IV. Zeros of a Function

A **zero** of a function is an x-value that results in an output of zero. That is, a is a zero of f if $f(a) = 0$.

Example 8 (Find the zeros of functions)

Find the zero(s) of each function.

(a) $f(x) = 2x^2 + x - 1$

(b) $g(x) = \dfrac{x-3}{2x-1}$

Example 9 (Find the values of x such that $f(x) = 0$)

Find all real values of x such that $f(x) = 0$.

(a) $f(x) = 4x + 1$

(b) $f(x) = x^3 - x$

Additional Notes

NAME: ______________________ EMAIL: ________________ @vols.utk.edu

Lesson 14: Graphs

Learning Objectives:

I. Identify the components of the Cartesian plane, plot points, and determine the location of points based on the information given.

II. Create a table of values for an equation in two variables and determine whether a point is on an equation's graph.

III. Find x- and y-intercepts graphically and algebraically.

IV. Determine whether a graph represents a function using the Vertical Line Test.

V. Evaluate a function graphically.

VI. Identify the domain and range of a function graphically.

VII. Identify the intervals on which a function is increasing, decreasing, or constant.

VIII. Identify the intervals on which a function is positive or negative.

I. Quadrants & Point Plotting

Ordered pairs of real numbers are represented by points in a plane. This plane is called the **Cartesian plane** and is formed by two real number lines intersecting at right angles.

- The ______________ is the horizontal real number line. The ______________ is the vertical real number line.

- The ______________ is the point where the two axes intersect.

- The two axes divide the plane into four parts called ______________ .

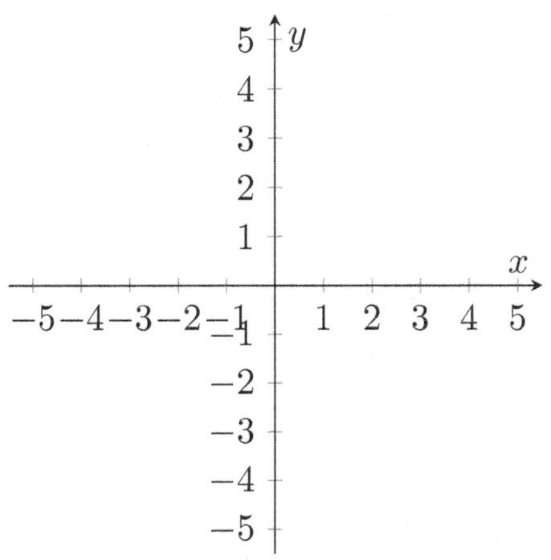

Each point in the plane corresponds to an **ordered pair** (x, y) of real numbers x and y, called the **coordinates** of the point. Plot (and label) the ordered pairs $(-2, 4)$ and $(3, -2)$ on the plane.

Example 1 (Determine the location of points based on the information given)

Determine the quadrant(s) in which (x, y) is located so that the condition(s) are satisfied.

(a) $x < 0$, $y < 0$

(b) $xy > 0$

II. Table of Values

The **graph of an equation** is the set of all points that are solutions of the equation. One way to graph an equation is to generate a set of ordered pairs that correspond to the equation. We call this *making a table of values.* Plotting the points from a table of values is not the best way to graph an equation because the points you choose for your table may not be the points that show the entire shape of the graph. However, generating a table of values is an important skill that will be used along with other skills to create graphs of equations.

Example 2 (Create a table of values for an equation)

Generate a table of values with at least four points for the equation $y = 3x + 1$.

x				
y				
(x, y)				

Example 3 (Create a table of values for an equation)

Generate a table of values with at least four points for the equation $y = x^2 - x + 1$.

x				
y				
(x, y)				

Example 4 (Determine whether a point is on an equation's graph)

Determine whether the points $(0, 3)$ and $(7, -4)$ lie on the graph of the equation $y = \sqrt{x + 9}$.
Hint: This is the same as asking if each point is in the table of values for the equation.

III. Intercepts of a Graph Visually & Algebraically

A point at which the graph of an equation meets the x-axis is called an ________________ .

A point at which the graph of an equation meets the y-axis is called a ________________ .

It is possible for a graph to have no intercepts, one intercept, or several intercepts.

Example 5 (Find x- and y-intercepts graphically)

Find the x- and y-intercepts of each graph shown below. List them as ordered pairs.

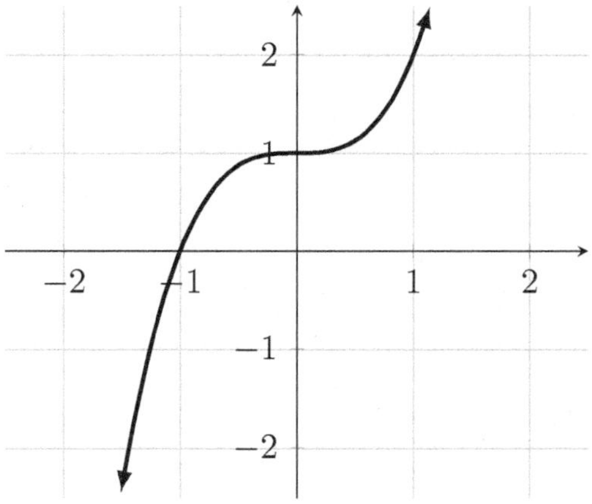

x-int: ____________________________

y-int: ____________________________

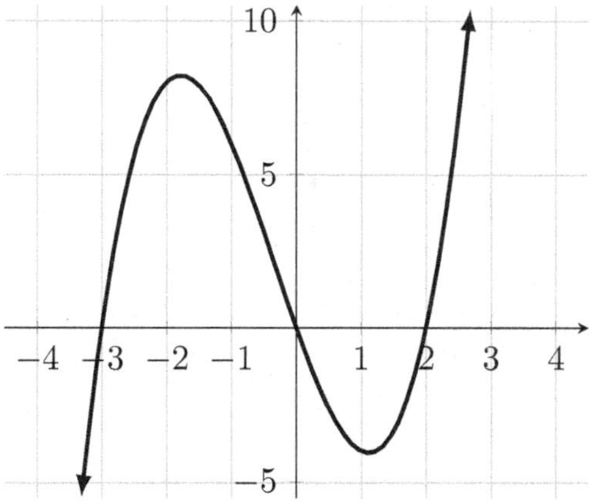

x-int: ____________________________

y-int: ____________________________

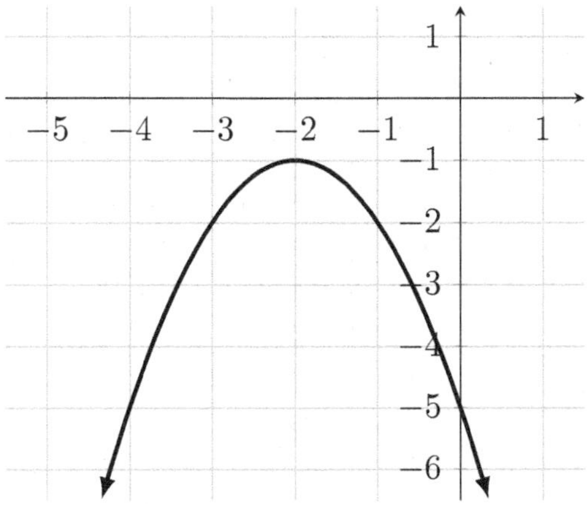

x-int: ____________________________

y-int: ____________________________

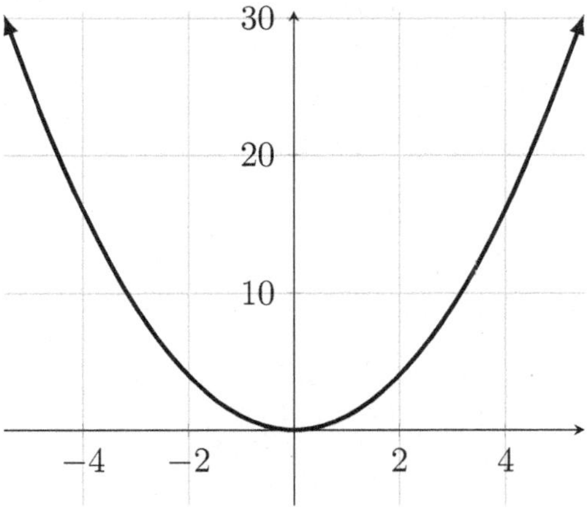

x-int: ____________________________

y-int: ____________________________

An x-intercept has the form __________ . A y-intercept has the form __________ . Always write the intercepts as **ordered pairs**.

Finding x- and y-intercepts algebraically involves solving a linear equation.

Example 6 (Find x- and y-intercepts algebraically)

Find the x- and y-intercepts for the equation $3x + 5y = 15$.

IV. The Vertical Line Test

The **Vertical Line Test** can be used to determine if a graph represents a ____________________ .

> **The Vertical Line Test**
> A set of points in a coordinate plane is the graph of y as a function of x if and only if no vertical line intersects the graph at more than one point.

Example 7 (Determine whether graphs represent functions)

Determine if the following are graphs of y as a function of x.

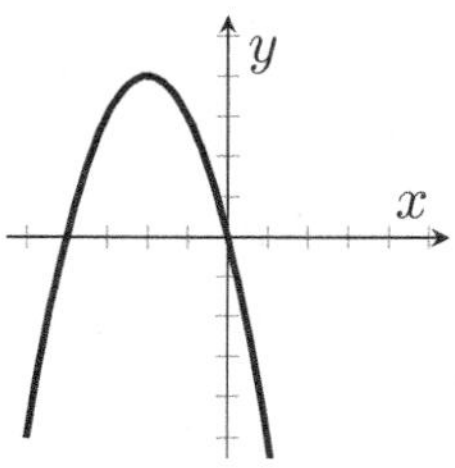

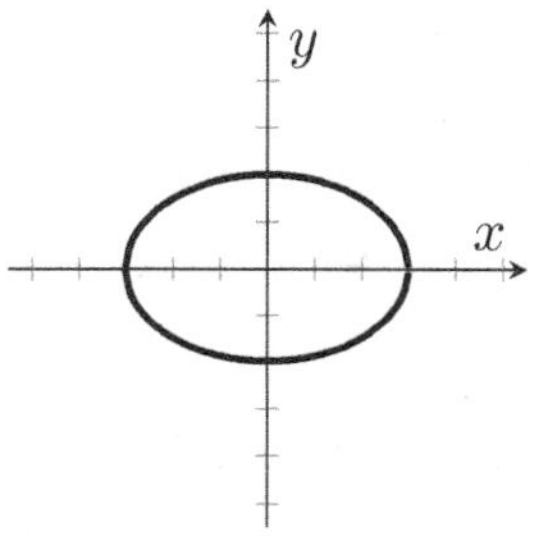

V. Evaluating Functions Graphically

A function can be evaluated using its graph. To find the value of $f(x)$ for a given value of x, locate the point on the graph corresponding to the x-value and determine the y-value of the point.

Example 8 (Evaluate a function graphically)

Find the following, given the graph of $f(x)$.

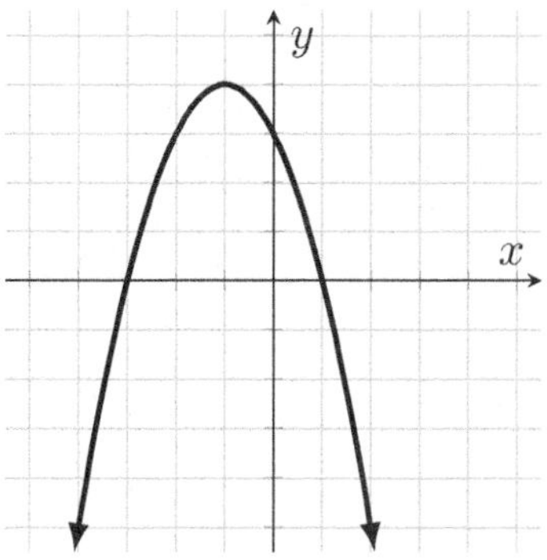

$f(-3) =$ __________

$f(-1) =$ __________

$f(0) =$ __________

$f(1) =$ __________

VI. Domain & Range From a Graph

You can tell the domain and range of a function by looking at its graph. Remember that the **domain** is the set of ________________ for which $f(x)$ is defined as a real number. Graphically, that means that the domain is the set of x-values for which there are corresponding points on the graph. Read the graph from ____________________ to determine which x-values are covered by the graph. The **range** is the set of ________________ for which there are corresponding points on the graph. Read the graph from ____________________ to determine which y-values are covered by the graph.

Example 9 (Domain and range of functions graphically)

Find the domain and range of each function shown.

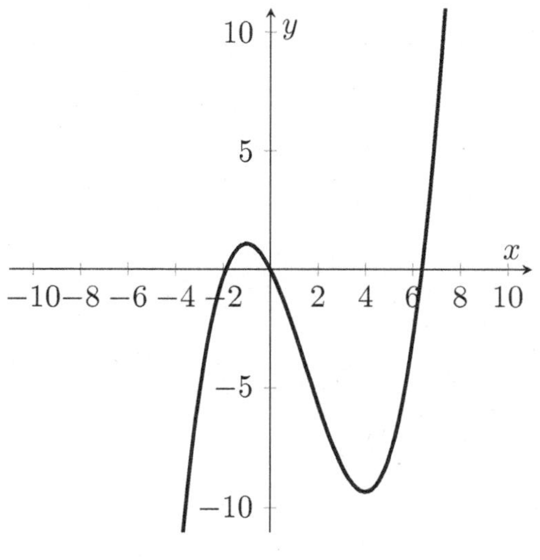

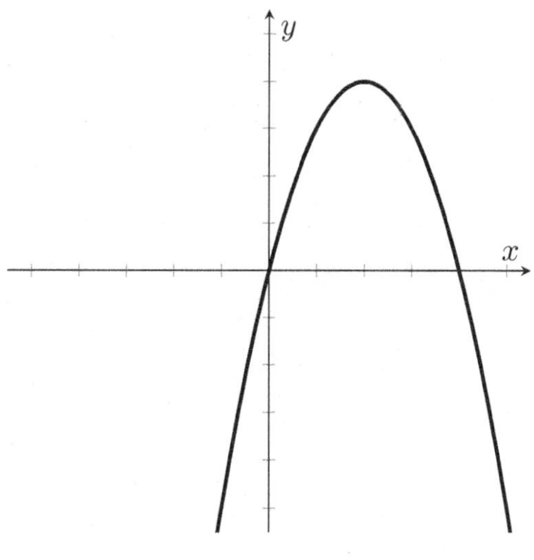

Domain: ________________________ Domain: ________________________

Range: ________________________ Range: ________________________

VII. Intervals on which a Function is Increasing, Decreasing, or Constant

To identify the (open) intervals on which the graph of a function is increasing, decreasing, or constant, read the graph from ________________ . As you go from left to right:

- If the y-values of the points get larger, the graph is ________________ .
- If the y-values of the points get smaller, the graph is ________________ .
- If the y-values of the points stay the same, the graph is ________________ .

Intervals on which the graph is increasing, decreasing, or constant are always listed in terms of the ________________ (even though you're looking at what's happening to the y-values), and the endpoints are **NOT** included (so use parentheses when writing the interval in interval notation).

When a function changes from ________________ to ________________ , it has a **relative maximum** (i.e., high spot) at that point. When it changes from ________________ to ________________ , it has a **relative minimum** (i.e., low spot) at that point.

Example 10 (Intervals on which a function is increasing, decreasing, or constant)

Use the graph below to list the domain and the range. List any intervals on which the function is increasing, decreasing, and constant. Also, label any relative maximums and/or relative minimums.

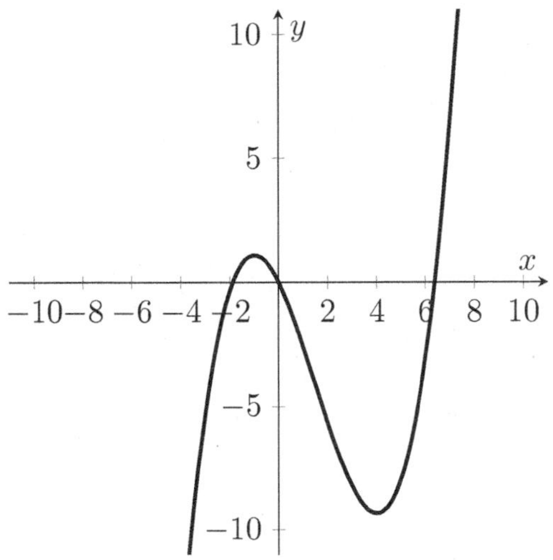

Domain: ________________

Range: ________________

Increasing: ________________

Decreasing: ________________

Constant: ________________

VIII. Intervals on which a Function is Positive or Negative

To identify the intervals on which the graph of a function is positive (i.e., $f(x) > 0$), look to see where the graph sits **above** the x-axis. To identify the intervals on which the graph of a function is negative (i.e., $f(x) < 0$), look to see where the graph sits **below** the x-axis. In either case, list the interval in terms of the x-values. Do not include the value(s) that make $f(x) = 0$.

Finding the intervals on which $f(x) \geq 0$ or $f(x) \leq 0$ is similar, except we include the value(s) that make $f(x) = 0$.

Example 11 (Intervals on which a function is positive or negative)

Use the graph below to solve the following equation and inequalities.

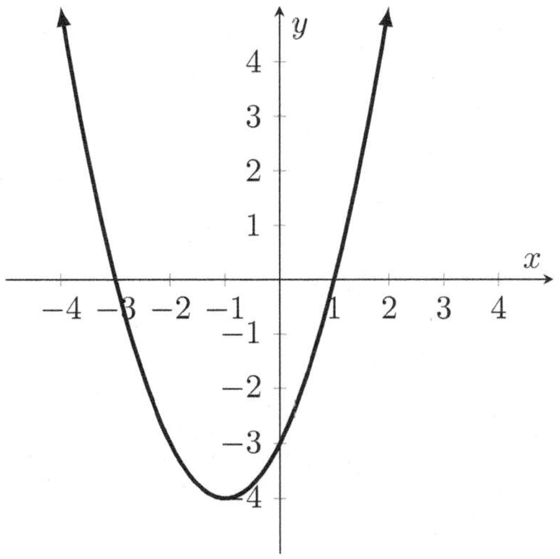

Solution(s) to $f(x) = 0$:

____

Solutions to $f(x) > 0$:

____

Solutions to $f(x) < 0$:

____

Example 12 (Intervals on which $f(x) \geq 0$ or $f(x) \leq 0$)

Use the graph below to solve the following equation and inequalities.

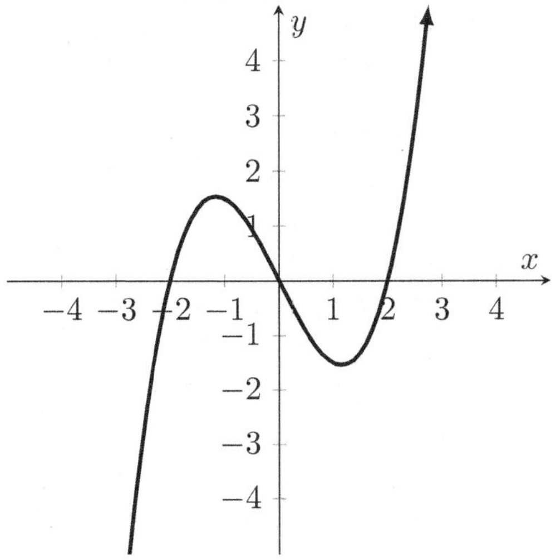

Solution(s) to $f(x) = 0$:

____

Solutions to $f(x) \geq 0$:

____

Solutions to $f(x) \leq 0$:

____

Additional Notes

NAME: ______________________ EMAIL: ________________ @vols.utk.edu

Lesson 15: Transformations of Common Functions

Learning Objectives:

I. Recognize the name, equation, and graph of the six common functions.

II. Transform the graph of a common function by shifting horizontally or vertically.

III. Transform the graph of a common function by reflecting across the x- or y-axis.

IV. Write the equation representing a described transformation of a common function.

I. Common Functions

You need to **memorize** the **graphs** of the common functions and the **reference points** for each common function. You also need to know the **name** of each function and its **equation**. From knowing the graphs of the common functions, you should be able to list the domain, range, and open intervals on which the function is increasing, decreasing, or constant.

Constant Function

Equation: ______________

Domain: __________ Range: __________

Increasing: __________ Decreasing: __________

Constant: __________

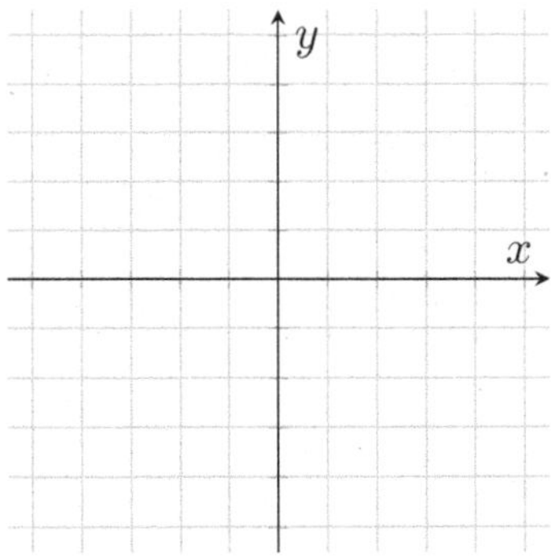

Identity Function

Equation: ______________

Reference Points: ____________________

Domain: __________ Range: __________

Increasing: __________ Decreasing: __________

Constant: __________

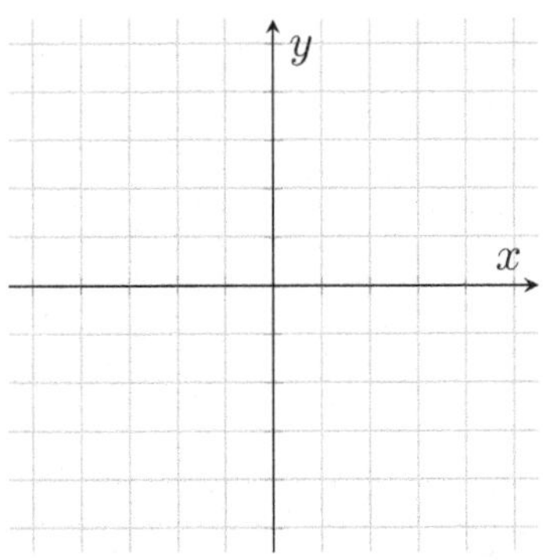

Absolute Value Function

Equation: ________________

Reference Points: ____________________

Domain: ____________ Range: ____________

Increasing: ____________ Decreasing: ____________

Constant: ____________

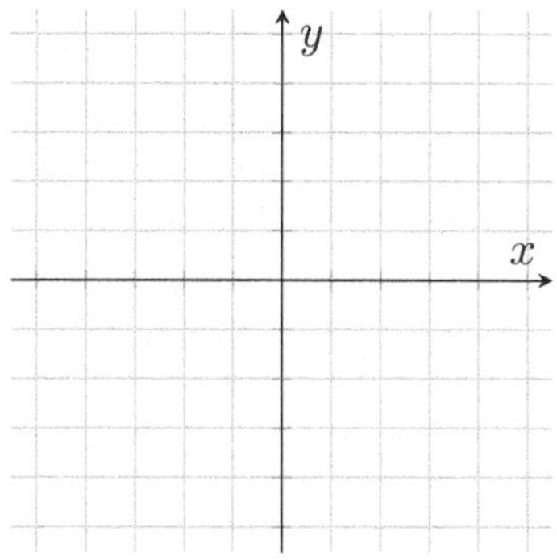

Square Root Function

Equation: ________________

Reference Points: ____________________

Domain: ____________ Range: ____________

Increasing: ____________ Decreasing: ____________

Constant: ____________

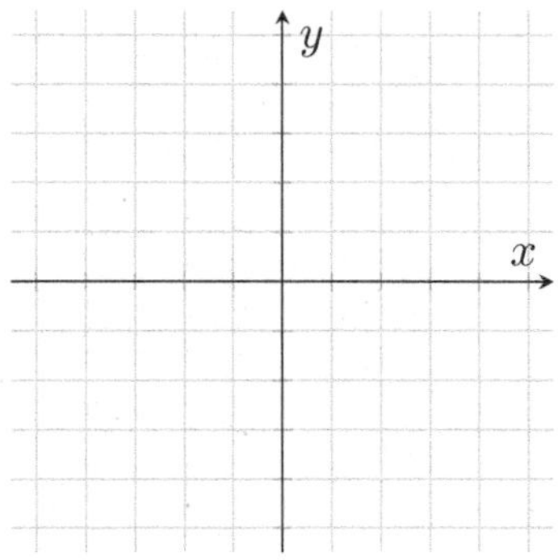

Quadratic Function

Equation: ________________

Reference Points: ____________________

Domain: ____________ Range: ____________

Increasing: ____________ Decreasing: ____________

Constant: ____________

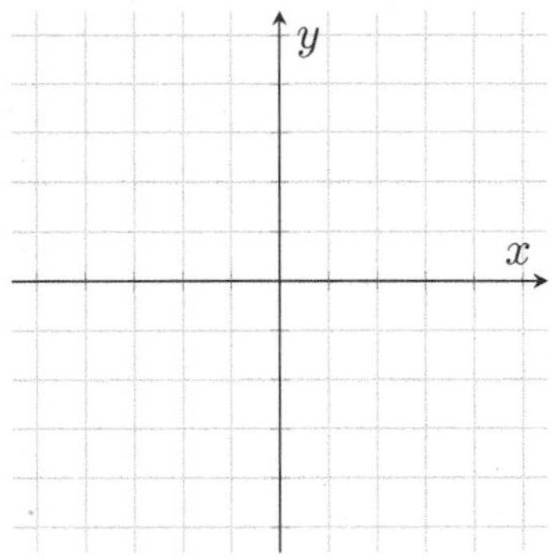

Cubic Function

Equation: ________________

Reference Points: ____________________

Domain: ____________ Range: ____________

Increasing: ____________ Decreasing: ____________

Constant: ____________

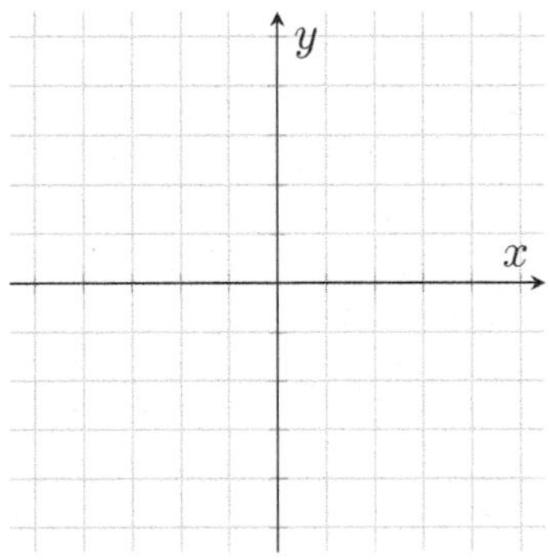

II. Shift Graphs of Common Functions

Let c be a positive real number. The following changes to the function $y = f(x)$ will produce the stated shifts in the graph. For our purposes, $y = f(x)$ will always be one of the common functions.

Horizontal Shifts

- $h(x) = f(x - c)$ produces a **horizontal** shift c units to the ______________ .
- $h(x) = f(x + c)$ produces a **horizontal** shift c units to the ______________ .

In either case, the positive real number is added or subtracted ______________ the common function.

Vertical Shifts

- $h(x) = f(x) - c$ produces a **vertical** shift c units ______________ .
- $h(x) = f(x) + c$ produces a **vertical** shift c units ______________ .

In either case, the positive real number is added or subtracted ______________ the common function.

Example 1 (Transform the graph of common functions by shifting)

Graph each function by shifting the graph of a common function.

(a) $y = |x - 2|$

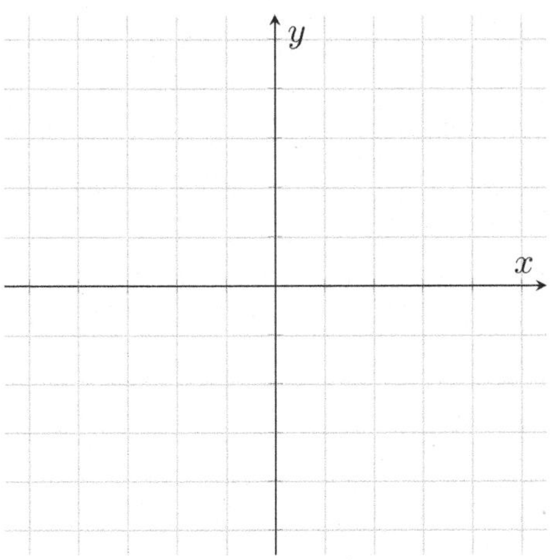

(b) $f(x) = \sqrt{x} + 2$

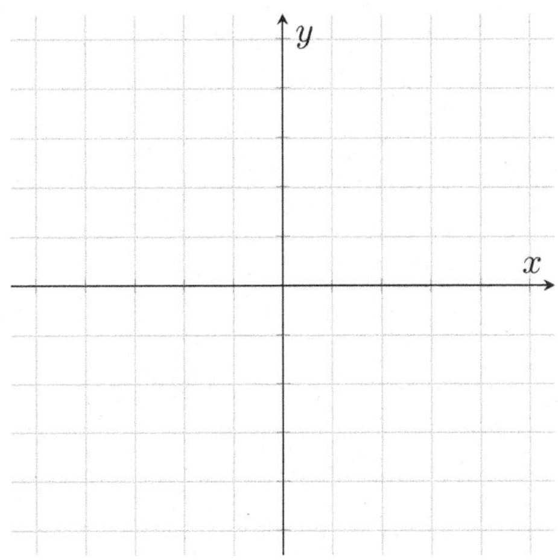

(c) $g(x) = (x+1)^2$

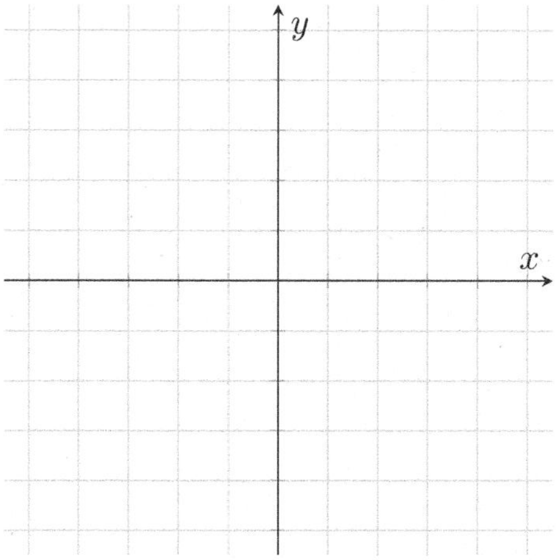

(d) $y = (x-2)^2 - 1$

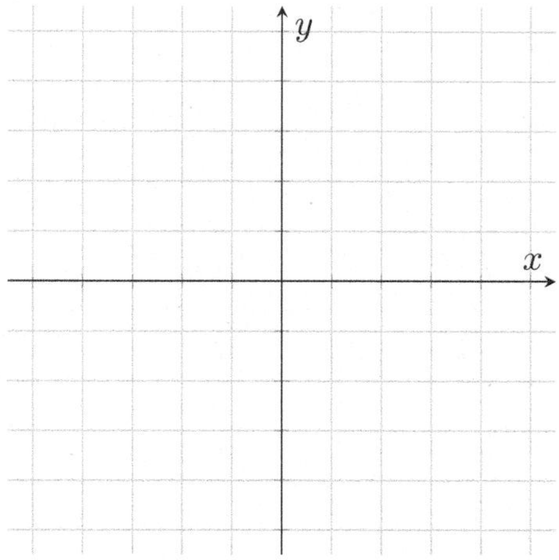

(e) $y = |x+1| - 2$

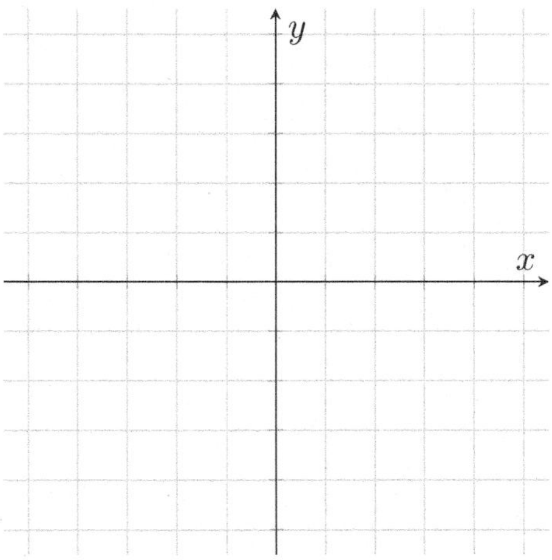

III. Reflect Graphs of Common Functions

The following changes in the function $y = f(x)$ will produce the stated reflections in the graph. For our purposes, $y = f(x)$ will always be one of the common functions.

Reflections

- $h(x) = f(-x)$ produces a reflection about the ______________ .

  In this case, the -1 is multiplied ____________ the common function.

- $h(x) = -f(x)$ produces a reflection about the ______________ .

  In this case, the -1 is multiplied ____________ the common function.

To do more than one transformation, use this order: horizontal, reflect, vertical (__________).

Example 2 (Transform the graph of common functions by reflecting and shifting)

Graph each function by transforming the graph of a common function.

(a) $f(x) = \sqrt{-x}$

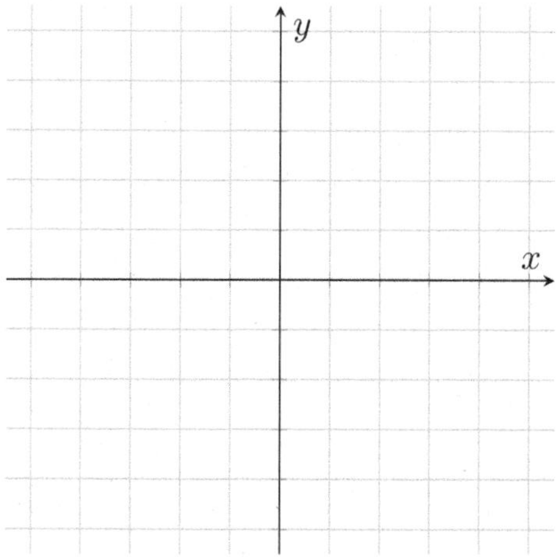

(b) $g(x) = -\sqrt{x}$

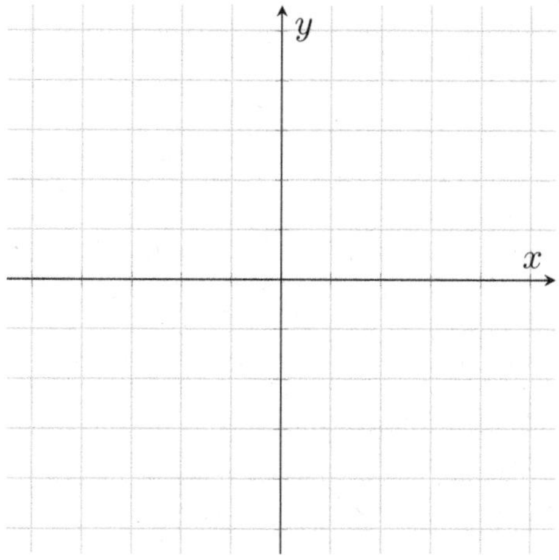

(c) $y = -x^3$

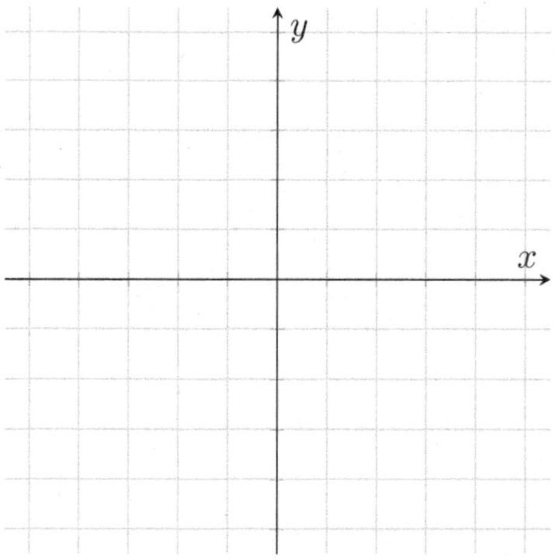

(d) $h(x) = -(x+3)^2$

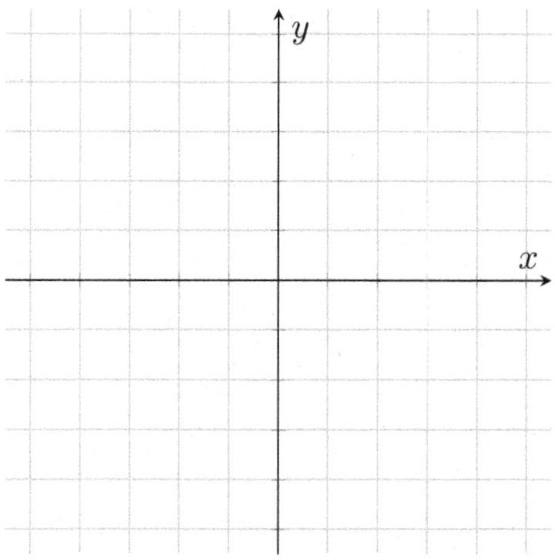

(e) $y = 2 - |x - 1|$

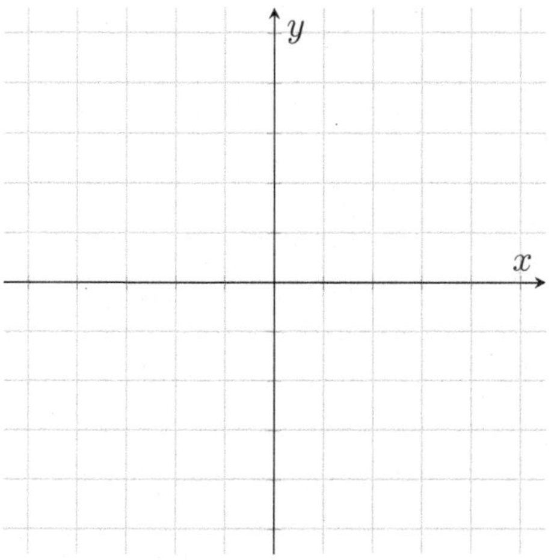

IV. Write Equations Representing Transformations

Example 3 (Write equations representing described transformations of comomn functions)

Write the equation for the function described.

(a) The quadratic function shifted to the right 3 units.

(b) The quadratic function shifted up 3 units.

(c) The cubic function shifted right 2 and reflected across the x-axis.

(d) The square root function shifted left 1 and up 4.

Additional Notes

Name: ______________________ Email: ________________ @vols.utk.edu

Lesson 16: Lines in the Plane

Learning Objectives:

I. Calculate the slope of a line algebraically using the slope formula. Recognize and analyze the equation of a line in slope-intercept form.

II. Sketch the graph of a line using its slope and y-intercept, or its x- and y-intercepts.

III. Write the equation of a line in slope-intercept form. Determine whether a line's equation indicates the graph will be diagonal, horizontal, or vertical.

IV. Describe the properties of parallel and perpendicular lines. Write the equation of a line, in slope-intercept form, that is parallel or perpendicular to a given line.

I. Slope & Slope-Intercept Form

Slope is a measure of the steepness of a line. Considering two points on the line, slope is the quotient of the vertical change between the two points and the horizontal change between the two points (i.e., "rise over run"). The formula for the slope of the line that goes through the points (x_1, y_1) and (x_2, y_2) is:

Positive Slope

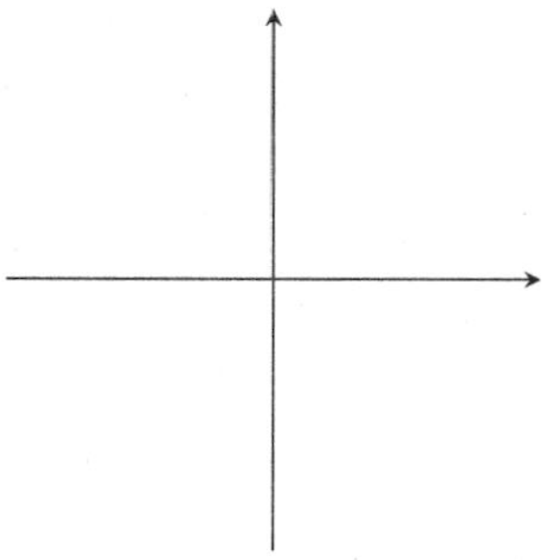

Negative Slope

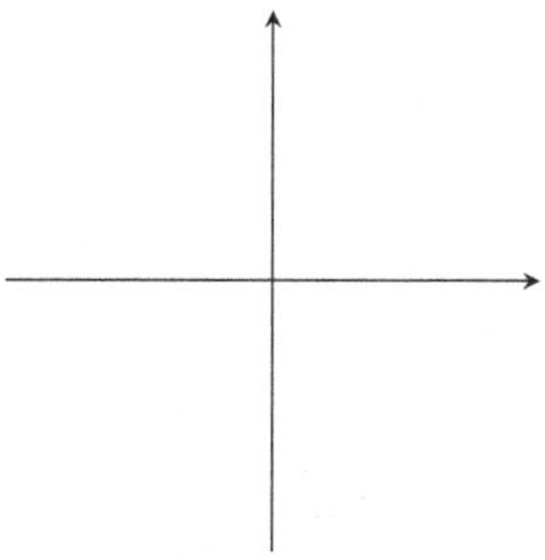

Zero Slope

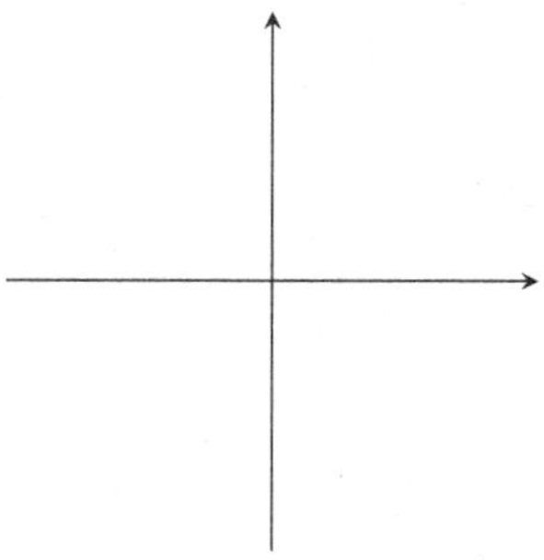

Undefined Slope

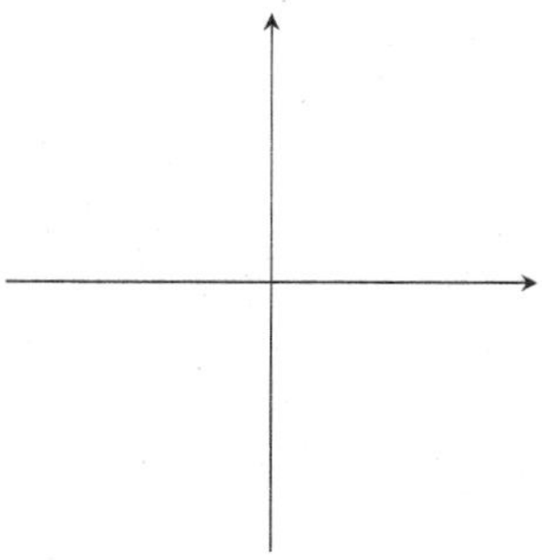

An equation of the form ______________ is a linear equation in **slope-intercept form.** Its graph is a line with **slope** ________ and **y-intercept** ____________ .

Example 1 (Calculate slope)

Find the slope of the line that goes through each pair of points.

(a) $(4, -6)$ and $(-4, 2)$

(b) $(5, -9)$ and $(-2, -9)$

(c) $(3, 2)$, and $(3, -4)$

II. Graph a Line

You can graph a line using any two points. If the equation is in slope-intercept form, it is often easiest to graph using the ____________________ . If the equation is in the form $Ax + By = C$, it is often easiest to graph using the ____________________ .

Example 2 (Graph lines using the slope and y-intercept)

Graph the line using the slope and y-intercept.

(a) $y = \frac{1}{3}x + 1$

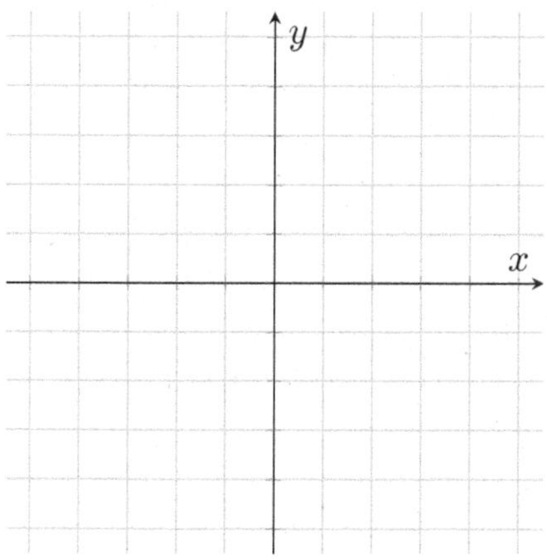

(b) $y = -3x - 1$

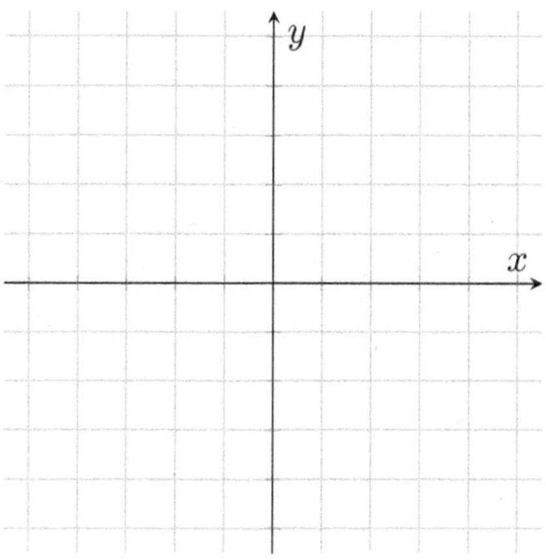

Example 3 (Graph lines using the x- and y-intercepts)

Graph the line $3x + y = 6$ using the x- and y-intercepts.

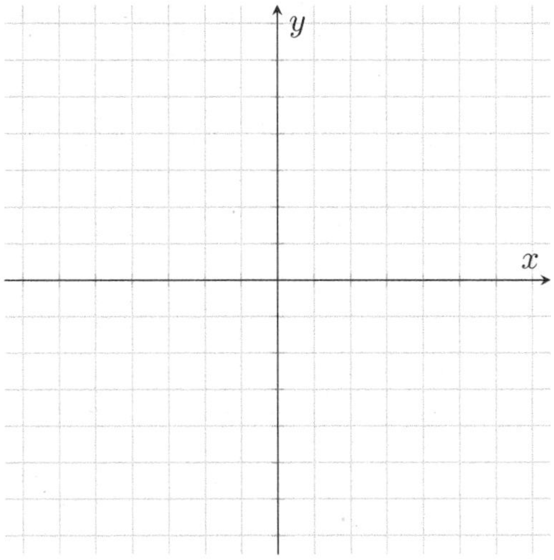

III. Write the Equation of a Line

To write the equation of a line, you must know the ________________ and the coordinates of one ________________ on the line. If you are only given the coordinates of two points, use them to find the slope.

There are two methods for writing the equation of a line:

1. **Use the Slope-Intercept formula**: Find the slope (if it is not given). Then substitute the x- and y-values of the point and the slope into the formula:

   $$\boxed{y = mx + b}$$

   Solve for b. Use m and b to write the equation of the line.

2. **Use the Point-Slope formula**: Find the slope (if it is not given). Then substitute the x- and y-values of the point and the slope into the formula:

   $$\boxed{y - y_1 = m(x - x_1)}$$

   The x- and y-values of the point replace the subscripted variables in the formula. Use algebra to rewrite the equation so that it is in slope-intercept form.

You may use the method that seems easier for you. You do not have to know both methods. **Always give your final answer in slope-intercept form.**

Example 4 (Write the equation of a line)

Write the slope-intercept form of the equation of the line with slope 8 that passes through the point $(-2, 8)$.

Example 5 (Write the equation of a line)

Write the slope-intercept form of the equation of the line that passes through the points $(-1, 3)$ and $(5, 1)$.

Vertical and Horizontal lines have equations with only one variable. The equation of the vertical line through any point with an x-coordinate of a is: ______________. The equation of the horizontal line through any point with a y-coordinate of b is: ______________.

Example 6 (Sketch the graphs of vertical and horizontal lines)

Sketch the graphs of the lines $x = 1$ and $y = -2$.

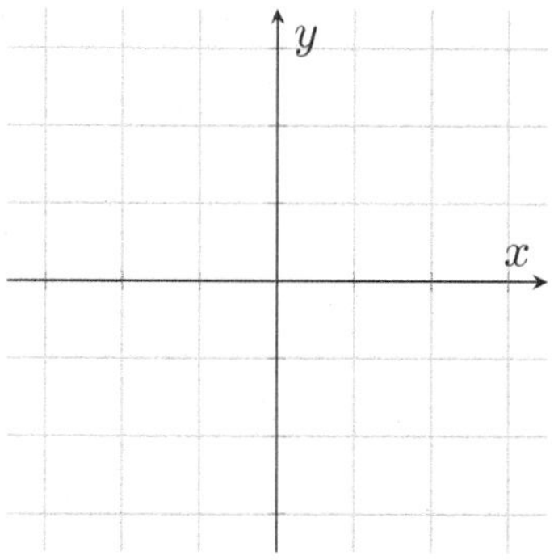

$x = 1$

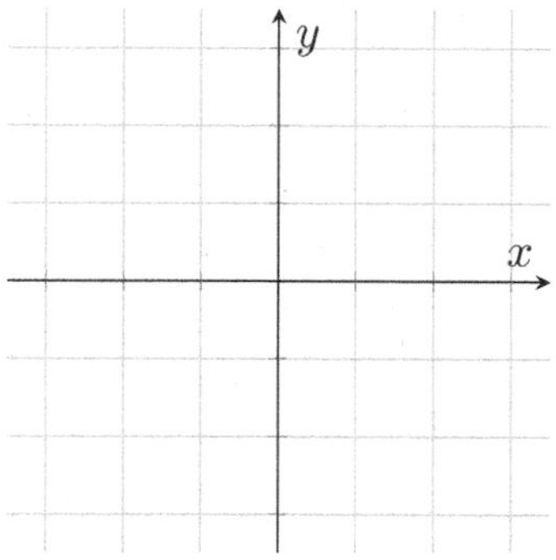

$y = -2$

Example 7 (Write the equations of vertical and horizontal lines)

Write the equations of the vertical and horizontal lines through the point $(4, -2)$.

IV. Parallel & Perpendicular Lines

- ____________________ lines have the **same slope**. Geometrically, if parallel lines are different from each other (i.e., if they are *distinct*), then they will never intersect. If parallel lines are the same (i.e., if they are *coincident*), then they will intersect at every point on the lines.

- ____________________ lines have slopes that are **negative reciprocals** (i.e., change the sign and flip). Geometrically, perpendicular lines intersect at one point and form a right angle.

Example 8 (Write equations of parallel and perpendicular lines)

Write the slope-intercept form of the equation of the line that passes through the point $(1, -3)$ and is (a) parallel to and (b) perpendicular to the line $3x + 5y = 1$.

Additional Notes

NAME: ______________________ EMAIL: ____________________ @vols.utk.edu

Lesson 17: Piecewise Functions

Learning Objectives:

I. Evaluate a piecewise function.

II. Sketch the graph of a linear piecewise function.

III. Analyze the properties of and write the equation of a graphed linear piecewise function.

I. Evaluate Piecewise Functions

Functions that are defined by two or more equations over a specified domain are known as

______________________ functions. To evaluate a piecewise function, first determine which expression should be used based on the specified domain for each piece of the function. Then evaluate using only that expression.

Example 1 (Evaluate a piecewise function)

Find the following, given $f(x) = \begin{cases} x+2 & \text{if } x \geq 0 \\ -x & \text{if } x < 0 \end{cases}$.

(a) $f(2) =$

(b) $f(0) =$

(c) $f(-5) =$

Example 2 (Evaluate a piecewise function)

Find the following, given $g(x) = \begin{cases} x^2+2 & \text{if } x < 2 \\ 5 & \text{if } 2 \leq x \leq 4 \\ 3x-1 & \text{if } x > 4 \end{cases}$.

(a) $g(0) =$

(b) $g(2) =$

(c) $g(3) =$

(d) $g(5) =$

II. Graph Piecewise Functions

To graph a piecewise function, make a small table of values for each piece. You must use the ________________ listed for that piece in your table even if the endpoint has a strict inequality (i.e., $<$ or $>$). Indicate whether each endpoint should be drawn using a ________________ dot (if the inequality is $\leq$ or $\geq$) or an ________________ dot (if the inequality is $<$ or $>$).

Example 3 (Graph piecewise functions)

Graph each piecewise function. List the domain and range, and any intervals on which the function is increasing, decreasing, and constant.

$$f(x) = \begin{cases} -2x & \text{if } x > 0 \\ x+1 & \text{if } x \leq 0 \end{cases}$$

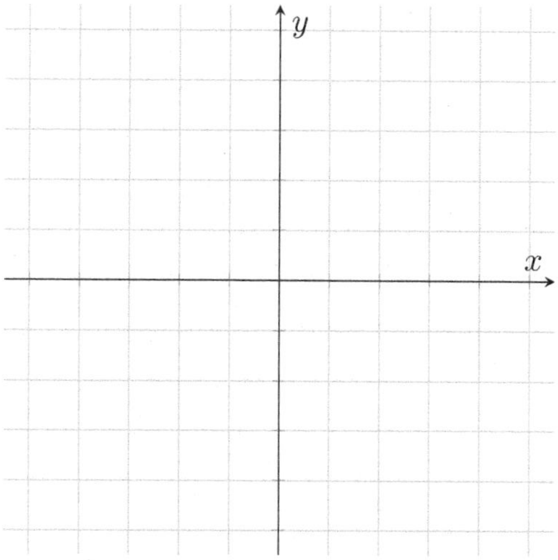

Domain: ____________________ Range: ____________________

Increasing: ____________ Decreasing: ____________ Constant: ____________

$$f(x) = \begin{cases} x+3 & \text{if } -4 < x \leq 0 \\ 2 & \text{if } 0 < x < 3 \\ -x & \text{if } x \geq 3 \end{cases}$$

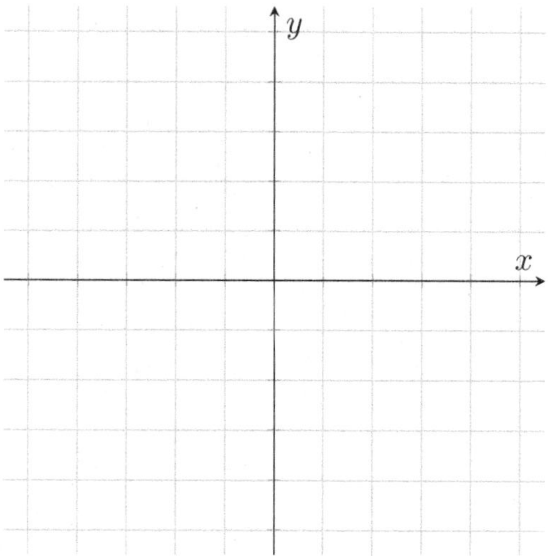

Domain: ____________________ Range: ____________________

Increasing: ____________ Decreasing: ____________ Constant: ____________

III. Analyze Graphs and Write Equations of Piecewise Functions

Example 4 (Analyze the graph of a piecewise function)

Use the graph below to list the domain and the range. Also, list any intervals on which the function is increasing, decreasing, and constant.

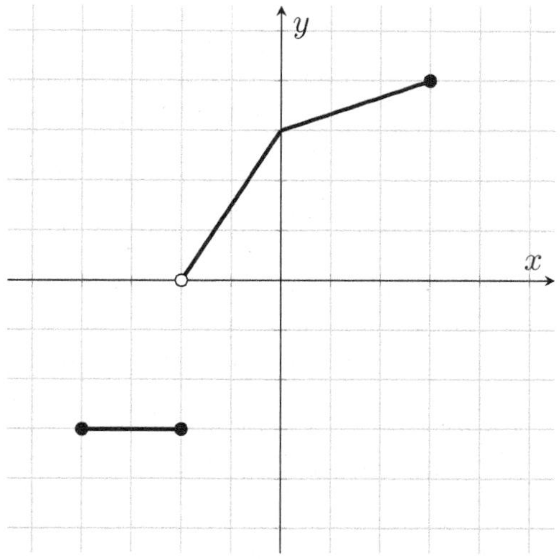

Domain: ____________________

Range: ____________________

Increasing: ____________________

Decreasing: ____________________

Constant: ____________________

Example 5 (Write the equations of piecewise functions)

Write the equation of the piecewise function shown in each graph.

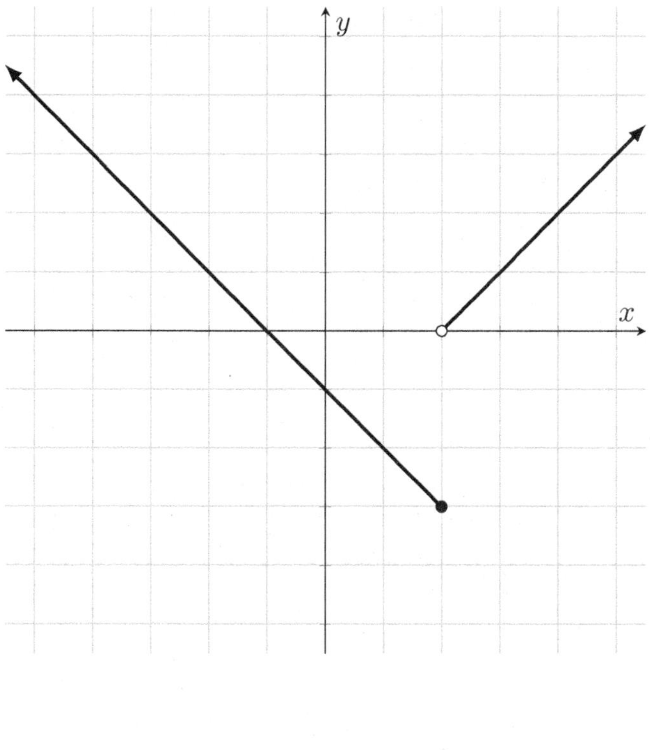

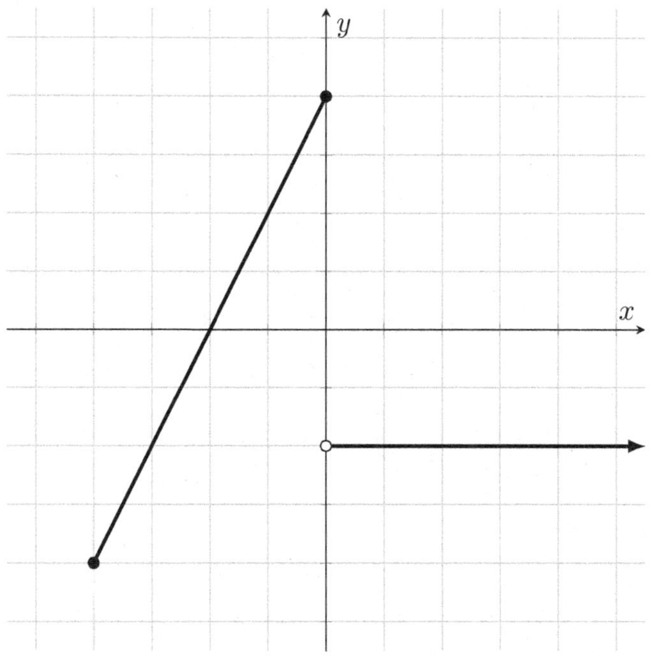
y
x

Additional Notes

NAME: ______________________ EMAIL: ________________ @vols.utk.edu

Lesson 18: Quadratic Functions

Learning Objectives:

I. Analyze the properties of a quadratic function and recognize how they affect the shape of its graph.

II. Find the vertex of a quadratic function.

III. Sketch the graph of a quadratic function using its vertex and x-intercepts.

IV. Sketch the graph of a quadratic function by transforming the graph of the common function $f(x) = x^2$.

I. Properties of Quadratic Functions

A quadratic function has **general form** ______________________ , with $a \neq 0$. The graph of a quadratic function is a ____________ . The ____________ is the point where the parabola changes from increasing to decreasing (or vice versa). It is also either a relative ____________ or a relative ____________ . The **axis of symmetry** is the ____________ line that goes through the vertex.

The equation for the axis of symmetry is:

The leading coefficient a reveals information about the shape and direction of the graph of the quadratic function $f(x) = ax^2 + bx + c$.

- When $a > 0$, the parabola opens ____________ .
- When $a < 0$, the parabola opens ____________ .
- When $0 < |a| < 1$, the parabola is ____________ than the common function $f(x) = x^2$.
- When $|a| > 1$, the parabola is ____________ than the common function $f(x) = x^2$.

Let's compare the graphs of $y = x^2$, $y = 2x^2$, and $y = \frac{1}{2}x^2$ to illustrate.

x	-1	0	1
$y = x^2$			

x	-1	0	1
$y = 2x^2$			

x	-1	0	1
$y = \frac{1}{2}x^2$			

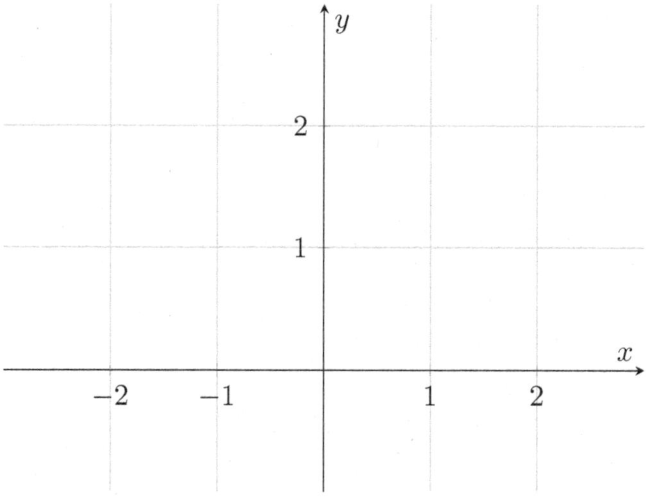

II. Finding the Vertex

To find the **vertex** of a parabola, use the formula in the box to the right.

The vertex is an ____________________ .

Hint: Don't memorize the formulas for both the x- and y−values. Just memorize the formula for the x-value and remember to plug that x-value into the function to find the y-value.

Example 1 (Find the vertices of quadratic functions)

Find the vertex of each quadratic function.

(a) $f(x) = -2x^2 - 8x - 4$

(b) $g(x) = x^2 - 5x + 6$

III. Graphing Using the Vertex and x-intercepts

Quadratic functions in general form can be graphed using the vertex and x-intercepts.

Example 2 (Sketch the graph of a quadratic function)

Sketch the graph of $g(x) = x^2 - 5x + 6$ using the vertex and x-intercepts. What is the axis of symmetry?

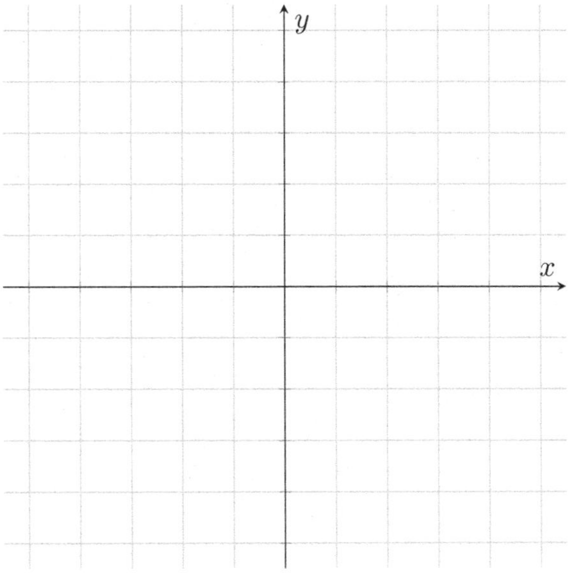

Does every quadratic function have x-intercepts? ____________

A quadratic function may have zero, one, or two x-intercepts.

Draw an example of each by shifting the common function $f(x) = x^2$. Write an equation for each graph.

One x-intercept:

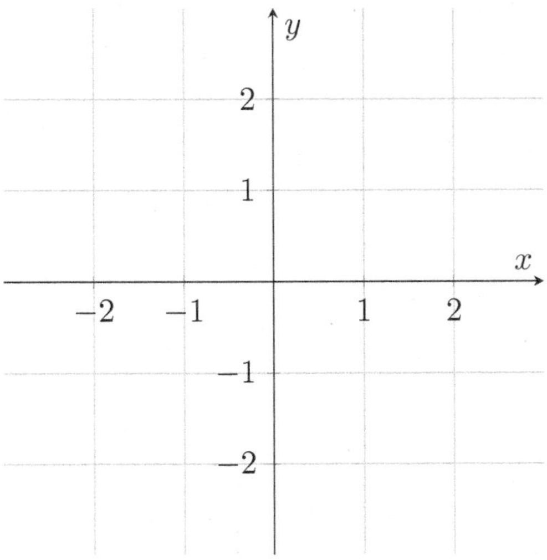

No x-intercepts:

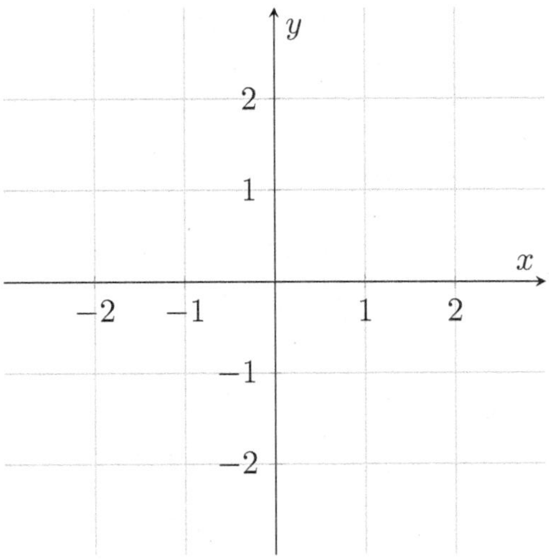

If the quadratic function does not have two x-intercepts to use to draw the graph, make a small table of values using points on both sides of the vertex to sketch the graph.

Example 3 (Sketch the graph of a quadratic function)

Sketch the graph of $f(x) = x^2 + 2x + 4$.

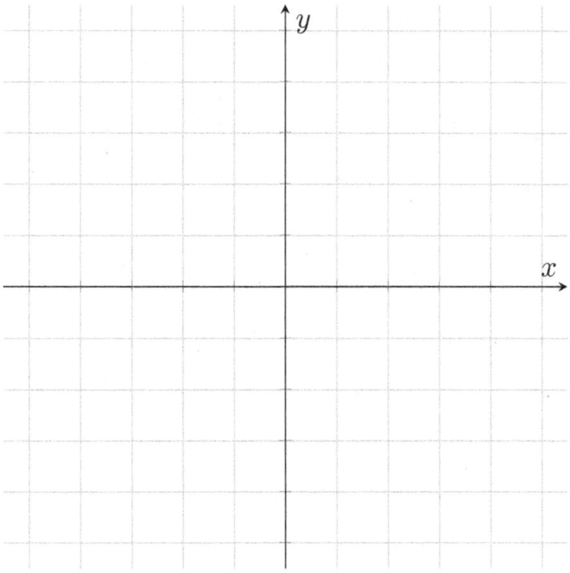

IV. Graphing Using Transformations

Some quadratic functions are given in **standard form** $f(x) = a(x-h)^2 + k$. To graph these, use the techniques you learned for transforming common functions. The sign of a tells you whether the parabola opens upward or downward. The h is the horizontal shift; the k is the vertical shift. If $|a| \neq 1$, the graph has been stretched or shrunk (i.e., is narrower or wider) by a factor of a. *You will not be asked to graph any quadratic functions involving stretching or shrinking.*

Example 4 (Sketch the graph of a quadratic function using transformations)

Sketch the graph of $f(x) = -(x+2)^2 + 1$.

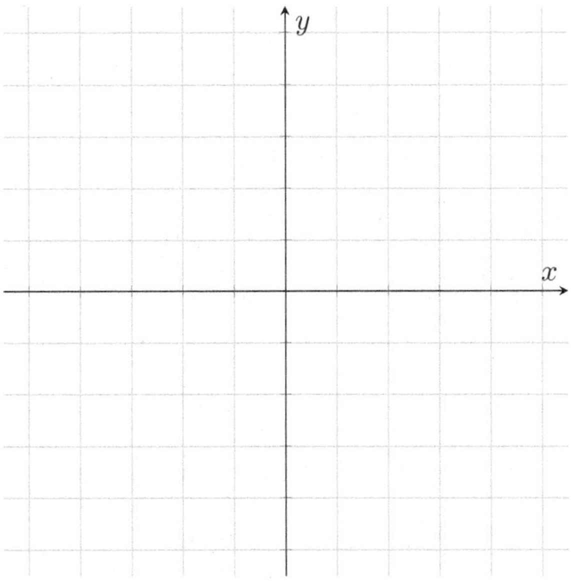

Example 5 (Find the vertex and x-intercepts of a quadratic function)

List the vertex and x-intercepts of $f(x) = -(x+2)^2 + 1$.
Hint: The vertex is found by shifting. Use extracting square roots to find the x-intercepts.

Additional Notes

NAME: ______________________________ EMAIL: _______________________ @vols.utk.edu

Lesson 19: Graphing Polynomial Functions

Learning Objectives:

I. Determine the maximum number of real zeros and turning points of the graph of a polynomial function.

II. Determine the end behavior of the graph of a polynomial function.

III. Calculate the multiplicities of the zeros of a polynomial function and analyze the effect of the multiplicity on the shape of the graph.

IV. Use the maximum number of turning points, zeros and their multiplicity, and end behavior to sketch an approximation of the graph of a polynomial function.

I. Turning Points

The graph of a polynomial function **of degree n** has at most __________________________ and at most __________________________ (or relative extrema). *("At most" means that many or fewer.)*

Example 1 (Maximum number of real zeros and turning points)

Find the maximum number of real zeros and the maximum number of turning points for the polynomial function $f(x) = 3x^5 - 2x^4 + 7x^2 - 5$.

Max Number of Real Zeros: Max Number of Turning Points:

II. Leading Coefficient Test

The **leading coefficient test** determines the end behavior of the graph. End behavior refers to what happens at each end of the graph. For example, both ends go up, both ends go down, etc.

	Leading Coefficient > 0	Leading Coefficient < 0
Even Degree		
Odd Degree		

Example 2 (End behavior)

Describe the end behavior of each function.

(a) $f(x) = -x^6 + 5x^4 - 3x^3 + 7$

(b) $g(x) = -3x^3 + x - 8$

III. Zeros & Multiplicity

The **zeros** of a polynomial tell you where the graph has ______________ . To find the zeros, set the function equal to ________ and solve by ______________ .

If $x = a$ is a zero, then ______________ is a factor. Some polynomials, when factored, have repeated factors (more than one factor that is the same). The number of times a factor is repeated is referred to as the ______________ of the zero. In general, $(x - a)^k$ yields a repeated zero of multiplicity ________ .

- If k is ____________ , the graph ______________ the x-axis at $x = a$.

- If k is ____________ , the graph ______________ the x-axis at $x = a$.

Example 3 (Compute multiplicities graphically)

Sketch the graph of each function and give the multiplicity of each zero.

(a) $f(x) = (x + 1)^2$

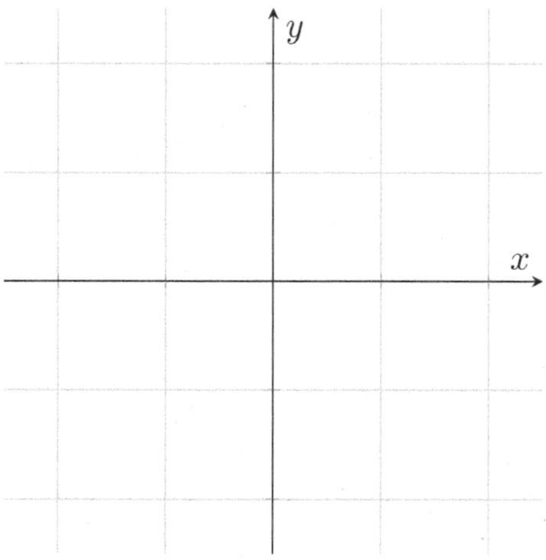

(b) $f(x) = (x + 1)^3$

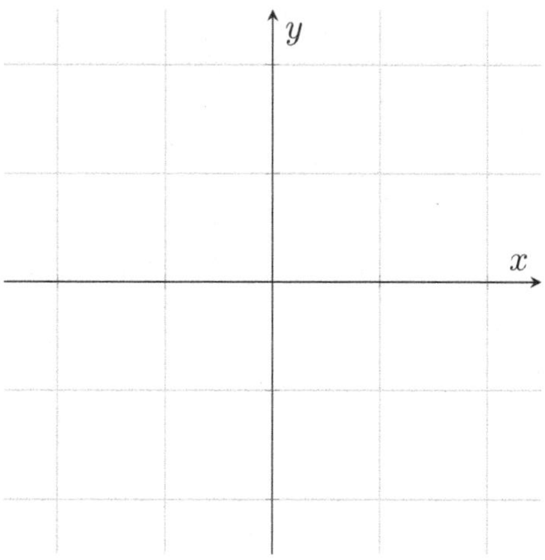

Example 4 (Compute multiplicities algebraically)

Find the zeros, give the multiplicity, and describe the behavior of the graph at each zero.

(a) $f(x) = 6x^4 - 33x^3 - 18x^2$

Zero	
Multiplicity	
Behavior	

(b) $g(x) = x(x+3)^4(x-4)^5$

Zero	
Multiplicity	
Behavior	

IV. Graph Polynomials

Example 5 (Sketch the graph of a polynomial function)

Sketch the graph of the function $f(x) = x^3 - 2x^2$ using your knowledge of the maximum number of turning points, end behavior, zeros/x-intercepts, and multiplicity. Plot a few additional points between the x-intercepts by making a table of values. Choose appropriate scales for your graph based on the table.

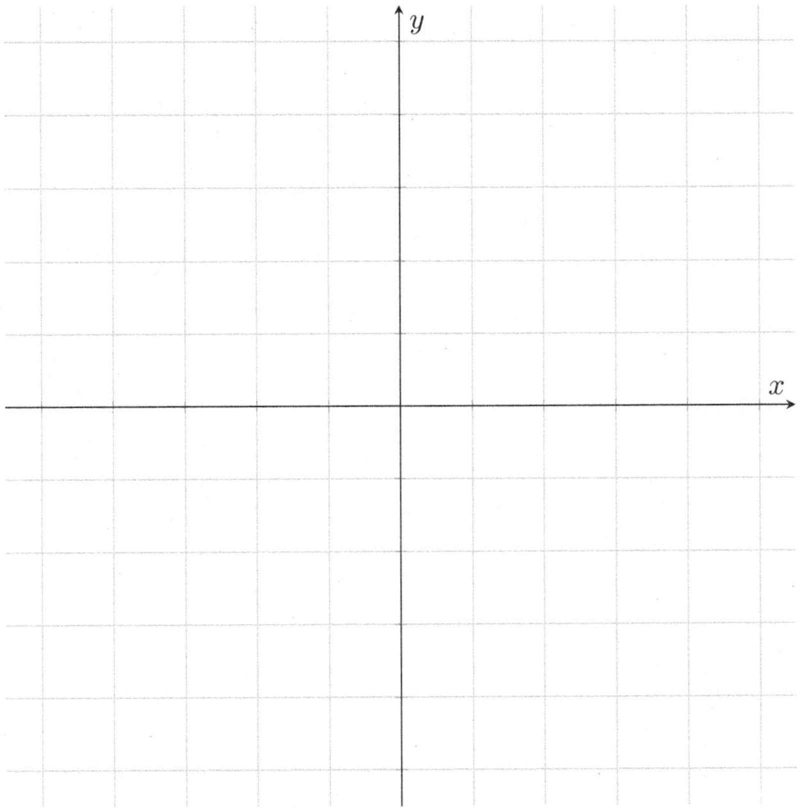

Additional Notes

NAME: ______________________ EMAIL: ____________________ @vols.utk.edu

Lesson 20: The Algebra of Functions

Learning Objectives:

I. Combine two functions using arithmetic operations.

II. Combine two functions using function composition.

III. Decompose a composite function into two functions.

I. Arithmetic Combinations of Functions

Let f and g be two functions with overlapping domains. The following arithmetic combinations of f and g are defined for all x-values common to both domains.

- Sum: $(f+g)(x) =$
- Difference: $(f-g)(x) =$
- Product: $(fg)(x) =$
- Quotient: $\left(\frac{f}{g}\right)(x) =$ (assuming that $g(x) \neq 0$)

Example 1 (Combine functions arithmetically)

Given $f(x) = x^2 + 3x$ and $g(x) = 3x + 4$, find the following.

(a) $(f+g)(x)$

(b) $(f+g)(2)$

(c) $(f-g)(x)$

(d) $(fg)(x)$

(e) $\left(\frac{f}{g}\right)(x)$

Note: You are expected to simplify your answers by distributing and collecting like terms (e.g., in parts a, c, and d). When finding the quotient of two functions, if factoring does not lead to a simplified answer (e.g., in part e), it is not required.

Example 2 (Multiply functions and give the domain)

Given $f(x) = \frac{1}{x}$ and $g(x) = \frac{x}{x+2}$, find and simplify $(fg)(x)$ and give its domain in words.

II. Function Composition

Composition is another way to combine two functions. When two functions are composed, one function becomes the ____________ for the other. The composition of the function f with g is $(f \circ g)(x) = f(g(x))$. The function g, the ____________ function, becomes the input for the function f, the ____________ function.

Example 3 (Compose functions)

Given $f(x) = x^2 + 3x$ and $g(x) = 3x + 4$, find the following.

(a) $(f \circ g)(x)$

(b) $(g \circ f)(x)$

(c) $(g \circ f)(3)$

(d) $(g \circ g)(x)$

Example 4 (Compose functions)

Given $f(x) = \sqrt{x-10}$ and $g(x) = x^2 + 1$, find the following.

(a) $(f \circ g)(x)$

(b) $(g \circ f)(x)$

Example 5 (Evaluate combinations of functions graphically)

Use the graphs of f and g to evaluate the functions.

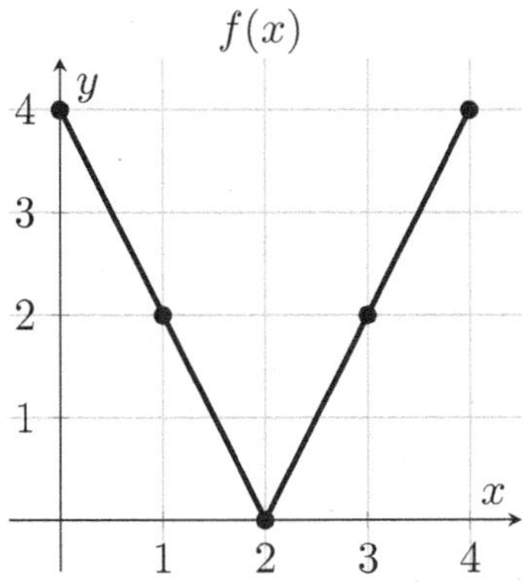

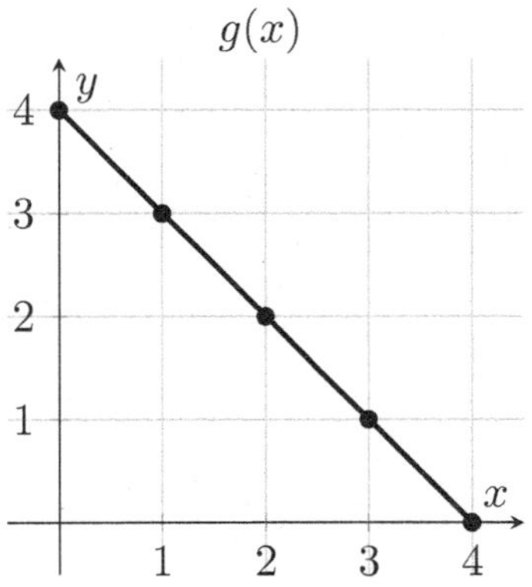

(a) $(f + g)(3)$

(b) $\left(\frac{f}{g}\right)(1)$

(c) $(f \circ g)(2)$

(d) $(g \circ f)(3)$

III. Function Decomposition

Decomposition refers to the process of figuring out which two functions could be composed to create a given function. Remember to think in terms of ____________ and ____________ functions.

Example 6 (Decompose composite functions)

Find two functions f and g such that $h(x) = f(g(x))$. (You may not use $f(x) = x$ or $g(x) = x$.)

(a) $h(x) = (x + 2)^4$

(b) $h(x) = \dfrac{6}{x^2 + 7}$

(c) $h(x) = \sqrt{3x + 4}$

Additional Notes

Unit 4

Exponential and Logarithmic Functions

In the last unit, we specifically examine exponential and logarithmic functions, which are essential for modeling growth, decay, and many real-world phenomena. In this unit, students will learn graph transformations for exponential and logarithmic functions, properties of logarithms, and how to solve both exponential and logarithmic equations.

NAME: ______________________ EMAIL: ____________________ @vols.utk.edu

Lesson 21: Exponential Functions and Their Graphs

Learning Objectives:

I. Sketch the graph and recognize the properties of a basic exponential function.

II. Sketch the graph of a transformation of a basic exponential function.

III. Sketch the graph and transformations of the natural exponential function.

I. Basic Exponential Function and its Graph

The **exponential function** with base a is denoted by $f(x) = a^x$, where $a > 0$, $a \neq 1$, and x is any real number. Its graph has the following properties:

Domain: __________ Range: __________ y-intercept: __________

Horizontal Asymptote: __________ (i.e. the __________)

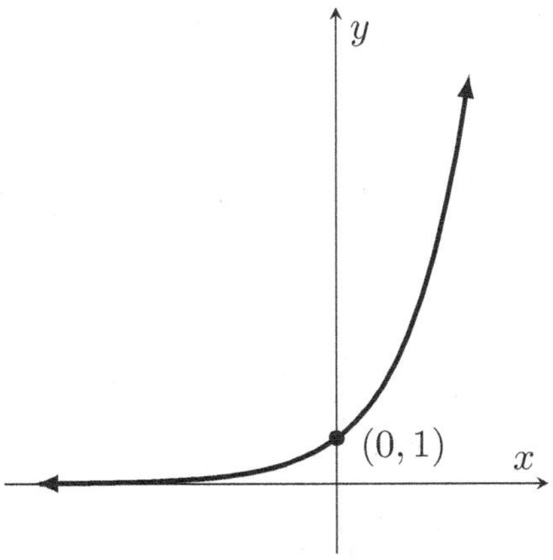

Increasing when $a > 1$

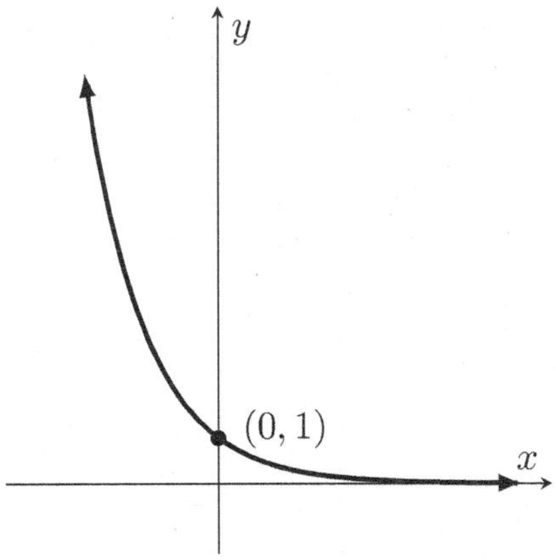

Decreasing when $0 < a < 1$

An exponential function written in the form $g(x) = a^{-x}$ can be rewritten (using exponent rules) as $g(x) = \left(\frac{1}{a}\right)^x$. To determine if functions of this form are increasing or decreasing, consider whether the base, $\frac{1}{a}$, is greater than 1 or between 0 and 1.

- When $a > 1$, the reciprocal, $\frac{1}{a}$, is between 0 and 1.

  So, the function $g(x) = a^{-x}$ is __________________ .

- When $0 < a < 1$, the reciprocal, $\frac{1}{a}$, is greater than 1.

  So, the function $g(x) = a^{-x}$ is __________________ .

To make a table of values for a basic exponential function, choose the numbers ____________ .

Example 1 (Graph basic exponential functions)

Sketch the graphs of $f(x) = 2^x$ and $g(x) = 2^{-x} = \left(\frac{1}{2}\right)^x$.

x			
$f(x)$			

x			
$g(x)$			

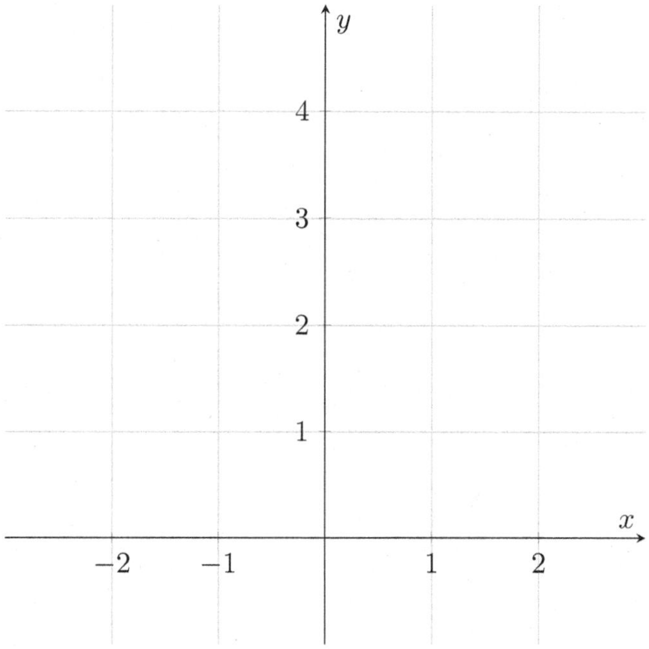

II. Transformations of $f(x) = a^x$

The transformations from Lesson 15 can be applied to functions of the form $f(x) = a^x$. As was the case in Lesson 15, c is a positive real number.

- $f(x + c) = a^{x+c}$ is a horizontal shift of $f(x)$ to the __________ c units.
- $f(x - c) = a^{x-c}$ is a horizontal shift of $f(x)$ to the __________ c units.
- $f(x) + c = a^x + c$ is a vertical shift of $f(x)$ __________ c units. (*Note: Asymptote moves.*)
- $f(x) - c = a^x - c$ is a vertical shift of $f(x)$ __________ c units. (*Note: Asymptote moves.*)
- $-f(x) = -a^x$ is a reflection of $f(x)$ across the __________ .
- $f(-x) = a^{-x}$ is a reflection of $f(x)$ across the __________ .

Example 2 (Graph a basic exponential function and its transformations)

Graph each of the following functions on the same coordinate axes. Graph the basic function using a small table of values with at least three points. Then use transformations to graph the other two functions. If the asymptote moves, include it as a dotted line.

(a) $f(x) = 3^x$

x			
$f(x)$			

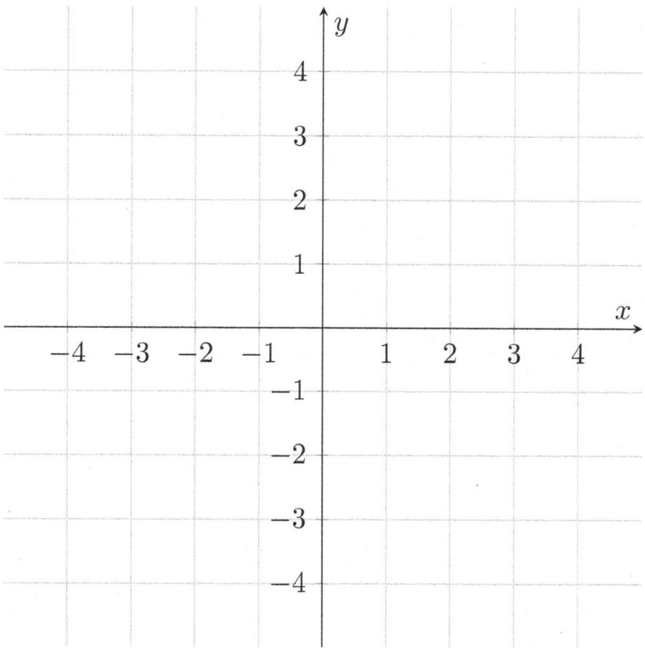

(b) $g(x) = 3^{x+2}$

(c) $h(x) = 3^{x+2} - 2$

III. The Natural Exponential Function

The **natural exponential** function, $f(x) = e^x$, has base e. The number e is irrational and is approximately equal to ________ .

Since $e > 1$, the natural exponential function is ____________________ . To graph the natural exponential function, make a table of values. It has the same shape as other increasing basic exponential functions. It can also be transformed like the other basic exponential functions.

Example 3 (Graph the natural exponential function and its transformations)

Graph each of the following functions on the same coordinate axes. Graph the basic function using a small table of values with at least three points. Then use transformations to graph the other two functions. If the asymptote moves, include it as a dotted line.

(a) $f(x) = e^x$

x			
$f(x)$			

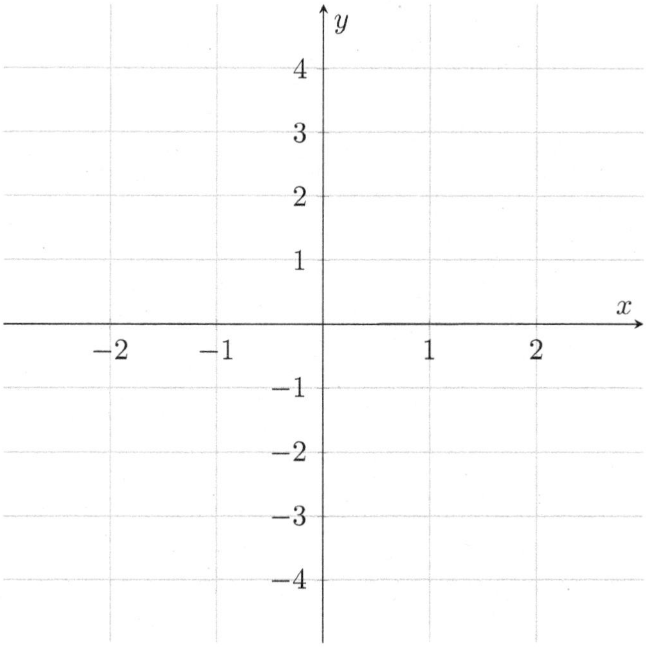

(b) $g(x) = -e^x$

(c) $h(x) = -e^x + 2$

Additional Notes

NAME: ______________________ EMAIL: ____________________ @vols.utk.edu

Lesson 22: Logarithmic Functions and Their Graphs

Learning Objectives:

I. Convert between logarithmic and exponential forms. Evaluate a logarithmic expression.

II. Define the common and natural logs. Use properties of logs to simplify a logarithmic expression.

III. Sketch the graph and transformations of a basic logarithmic function.

IV. Find the domain of a logarithmic function.

I. Basic Logarithmic Function

The **logarithmic function with base** a is denoted by $f(x) = \log_a x$, where $x > 0$, $a > 0$, and $a \neq 1$. The relationship between the logarithmic function and the exponential function is as follows:

$$y = \log_a x \quad \text{if and only if} \quad x = a^y$$

Example 1 (Convert between log and exponential forms)

Complete the tables by converting to log form or exponential form.

Log Form	Exponential Form
$\log_3 81 = 4$	
$\log_{10} \frac{1}{100} = -2$	
$\log_2 32 = 5$	

Log Form	Exponential Form
	$5^2 = 25$
	$16^{3/2} = 64$
	$10^{-1} = 0.10$

Example 2 (Evaluate logarithmic expressions)

Evaluate each logarithmic expression.

(a) $\log_2 16$ (b) $\log_2 0.25$ (c) $\log_{10} 1,000$

II. Common Log, Natural Log, and Properties

The two logs we use most often are the ____________________ (base 10), denoted $\log x$, and the ____________________ (base e), denoted $\ln x$. In other words, the notation $\log x$ means $\log_{10} x$, and $\ln x$ means $\log_e x$.

Example 3 (Evaluate logarithmic expressions)

Evaluate each logarithmic expression.

(a) $\log 100$ (b) $\log \frac{1}{10}$ (c) $\ln e^5$ (d) $\ln \frac{1}{e^2}$

Properties of Logs:

- $\log_a 1 =$ ____________
- $\log_a a =$ ____________
- $\log_a a^x =$ ____________
- $a^{\log_a x} =$ ____________
- One-to-One Property: if $\log_a x = \log_a y$, then $x = y$.

Example 4 (Log properties for the natural logarithm)

Complete the log properties for the natural logarithm.

- $\ln 1 =$ ____________
- $\ln e^x =$ ____________
- $\ln e =$ ____________
- $e^{\ln x} =$ ____________
- One-to-One Property: if $\ln x = \ln y$, then ____________ .

Example 5 (Use log properties to solve equations)

Use log properties to solve each equation.

(a) $\log_7 x = \log_7 12$

(b) $\log_4 1 = x$

(c) $e^{\ln 5} = x$

(d) $\log_6 6^x = 17$

III. Graph and Transformations

The properties of the graph of the basic log function, $f(x) = \log_a x$ where $x > 0$, $a > 0$, and $a \neq 1$, are as follows.

Domain: __________ Range: __________ x-intercept: __________

Vertical Asymptote: __________ (i.e. the __________)

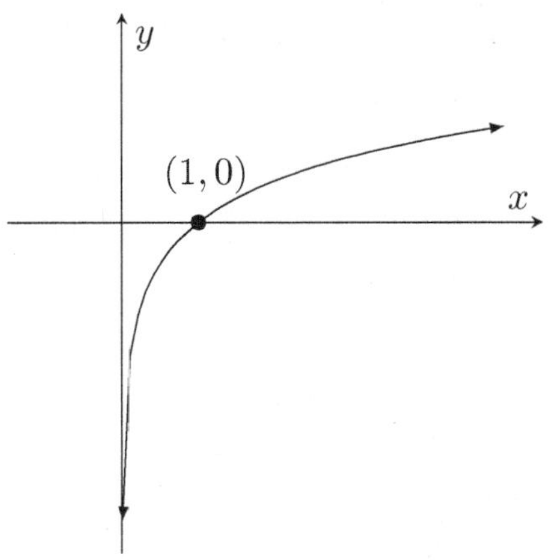

Increasing when $a > 1$

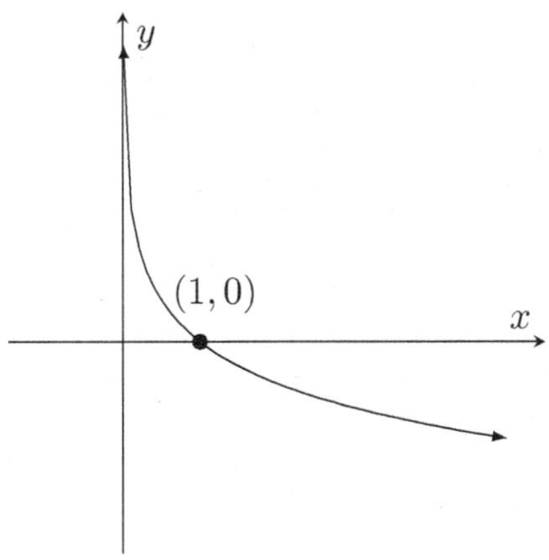

Decreasing when $0 < a < 1$

To make a table of values for a basic log function, choose the __________ of the base, the number ______ , and the __________ itself. Transformations of the basic graph work just like in previous sections. A __________ shift causes the vertical asymptote to move.

Example 6 (Graph a basic log function)

Sketch the graph of $f(x) = \log_3 x$ by making a small table of values with at least three points.

x			
$\log_3 x$			

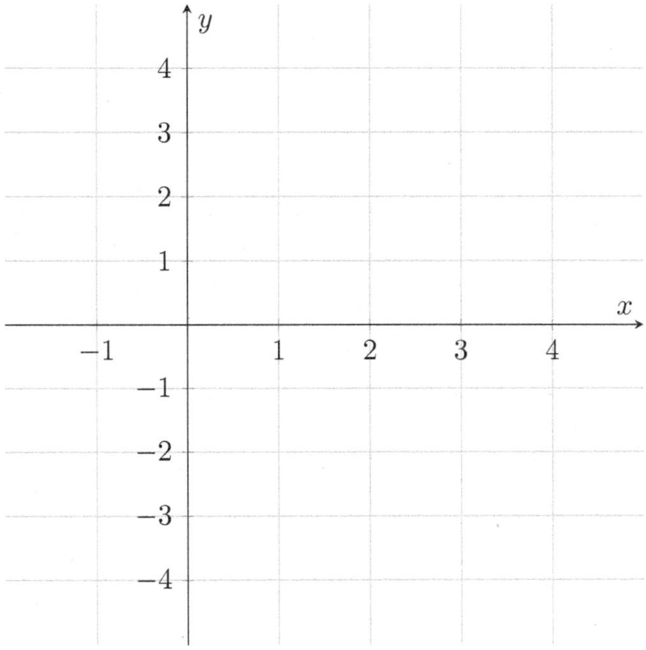

Example 7 (Graph a basic log function and its transformations)

Graph each of the following functions on the same coordinate axes. Graph the basic function using a small table of values with at least three points. Then use transformations to graph the other two functions. If the asymptote moves, include it as a dotted line.

(a) $y = \log_2 x$

x			
$\log_2 x$			

(b) $y = \log_2(x + 1)$

(c) $y = \log_2(x + 1) - 2$

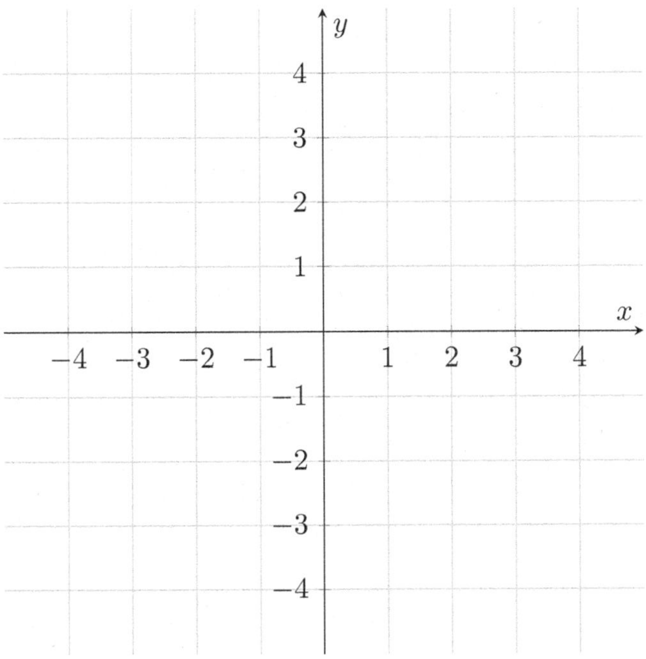

IV. Domain of a Logarithmic Function

The logarithm is only defined for ________________ values. You cannot take the log of 0 or the log of a negative number. The log of anything less than or equal to 0 is ________________ .

When finding the domain of a logarithmic function, you need to determine which values make the expression in the log statement ________________ .

Example 8 (Domain of a logarithmic function)

Find the domain of the function $g(x) = \log(1 - 3x)$. Write the answer in interval notation.

For some log functions, the domain can be determined by considering how the graph of a basic log function has been shifted.

Example 9 (Domain of logarithmic functions)

For each function, find the domain, vertical asymptote, and x-intercept.

(a) $f(x) = \ln(x - 3)$

(b) $g(x) = \log_5(x + 2)$

Additional Notes

NAME: ______________________ EMAIL: ____________________ @vols.utk.edu

Lesson 23: Properties of Logarithms

Learning Objectives:

I. Expand or condense a logarithmic expression using the properties of logarithms.

II. Evaluate a logarithmic expression without using a calculator.

I. Condense and Expand Properties of Logs

To **condense** a logarithmic expression means to write it as a ______________ logarithm with a coefficient of ________ .

To **expand** a logarithmic expression means to write the log as a sum or difference of ______________ logs (i.e., logs that cannot be expanded further).

-
Condensed Form	=	Expanded Form
$\log_a(uv)$	=	$\log_a(u) + \log_a(v)$

The log of a ______________ can be written as the ______________ of the logs of each factor.

-
Condensed Form	=	Expanded Form
$\log_a\left(\frac{u}{v}\right)$	=	$\log_a(u) - \log_a(v)$

The log of a ______________ can be written as the log of the ______________ minus the log of the ______________ .

-
Condensed Form	=	Expanded Form
$\log_a(u^n)$	=	$n\log_a(u)$

The log of something to a ______________ can be written as the power ______________ the log. *(Bring the power to the front.)*

Properties of Logs:

$\log_a(uv) = \log_a(u) + \log_a(v)$	$\log_a\left(\frac{u}{v}\right) = \log_a(u) - \log_a(v)$	$\log_a(u^n) = n\log_a(u)$

Example 1 (Expand logarithmic expressions)

Expand each logarithmic expression.

(a) $\log(3x^4y^5)$

(b) $\ln\left(\frac{\sqrt{x+2}}{y^4}\right)$

(c) $\log_2\left(\sqrt[4]{\frac{a^3}{b}}\right)$

(d) $\log\left(\frac{4x}{7y}\right)$

Example 2 (Condense logarithmic expressions)

Condense each logarithmic expression.

(a) $3\log(x) - 4\log(y) + \frac{1}{2}\log(z)$

(b) $\frac{1}{5}\left(3\ln(x) - 2\ln(y) - \ln(w+4)\right)$

(c) $\log(x+1) + \log(x-3)$

II. Evaluate Logs without a Calculator

The properties of logs can sometimes be used to evaluate logs that would otherwise be difficult to compute without a calculator.

Example 3 (Evaluate logarithmic expressions)

Evaluate each log without a calculator.

(a) $\log_3(81^{-0.2})$

(b) $2\ln(e^6) - \ln(e^5)$

(c) $\log_4(2) + \log_4(32)$

Additional Notes

NAME: ______________________ EMAIL: ______________________ @vols.utk.edu

Lesson 24: Exponential and Logarithmic Equations

Learning Objectives:

I. Solve an exponential equation.

II. Solve a logarithmic equation.

I. Solving Exponential Equations

An exponential equation is an equation that has the ____________ in an ____________ .

The **one-to-one property** for exponential equations states that if you can write both sides of the equation with the same ____________ , you can set the exponents ____________ to each other.

Solving Exponential Equations - Type A: (the equation can be written so that both sides have the same base)

- Rewrite both sides so that they have the same base.
- Use the one-to-one property for exponents to solve.

Example 1 (Solve exponential equations of type A)

Solve each equation.

(a) $2^{x-4} = 8$

(b) $\left(\frac{1}{4}\right)^{x+1} = 64$

(c) $e^{2x-3} = e^2$

(d) $10^{x+2} = \frac{1}{100}$

Solving Exponential Equations - Type B: (the equation cannot be written so that both sides have the same base)

- Isolate the exponential expression. (Get the part with the variable all by itself on one side of the equation.)
- Take the log of both sides. *(Choose the base of the log so that it matches the base of the exponential expression.)*
- Use log properties to solve.

Example 2 (Solve exponential equations of type B)

Solve each equation.

(a) $5e^{3x} = 30$

(b) $3e^{x+2} - 4 = 19$

(c) $5^x = 7$

(d) $2(4^x - 1) = 10$

II. Solving Logarithmic Equations

A logarithmic equation is an equation that has the ____________ in a ____________.

The **one-to-one property** for logarithmic equations states that if you can write both sides of the equation as a ____________ logarithm with the same base, you can set the ____________ (i.e., the stuff inside the logs) equal to each other. To get both sides of the equation written as a single log, you may have to ____________ the logs.

Solving Logarithmic Equations - Type A: (every term on both sides contains a logarithmic expression)

- Condense both sides of the equation.
- Use the one-to-one property of logarithms.
- Solve the resulting equation. (It may be linear, quadratic, rational, etc.)
- Check for extraneous solutions. (Extraneous solutions are solutions that are not in the domain of the logarithmic expression(s) in the original equation.)

Example 3 (Solve logarithmic equations of type A)

Solve each equation.

(a) $\ln(2x + 4) = \ln(3x)$

(b) $\log(x + 5) = \log(x - 1) - \log(x + 1)$

(c) $\log(x) + \log(x + 1) = \log(x + 5)$

(d) $2 \ln x = \ln(2x - 1)$

Solving Logarithmic Equations - Type B: (some of the terms in the equation are logarithmic expressions, but some terms are constant)

- Rewrite the equation so that all of the logs are on one side of the equation and the other side is constant.
- Condense the logs.
- Convert the equation to exponential form.
- Solve the resulting equation.
- Check for extraneous solutions.

Example 4 (Solve logarithmic equations of type B)

Solve each equation.

(a) $4\log_2(x+1) = 12$

(b) $\ln(5x) = 1$

(c) $\ln\sqrt{x+5} = 1$

(d) $\log(x+1) + \log(x+4) = 1$

Additional Notes